HOW TO RUN A LATHE

PRICE 50 CENTS

SOUTH BEND LATHE WORKS

SOUTH BEND, IND., U.S.A.

ISBN 978-1-60386-467-1

CONTENTS

How to Run a Lathe

FOR THE BEGINNER

How to Erect, Care For and Operate
a Screw Cutting Engine Lathe

REVISED EDITION NO. 15

O'BRIEN BROTHERS
J. J. O'Brien M. W. O'Brien

We have changed our name from South Bend Machine Tool Company to that of SOUTH BEND LATHE WORKS, to indicate more clearly the line of our manufacture

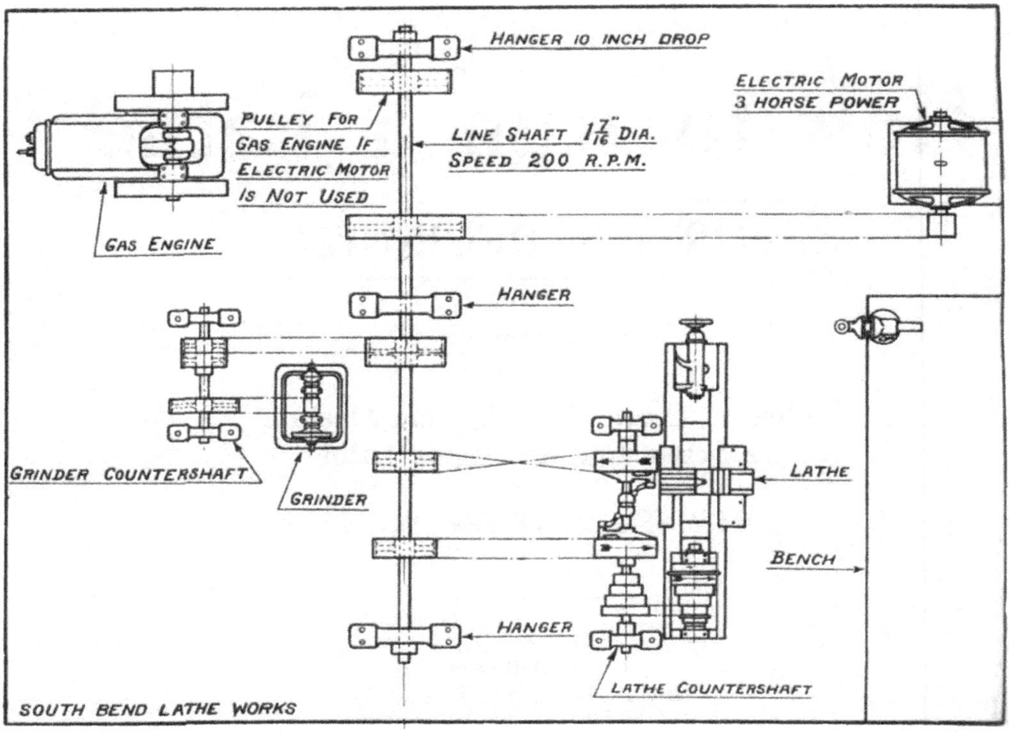
HANGER 10 INCH DROP
ELECTRIC MOTOR
3 HORSE POWER
PULLEY FOR
GAS ENGINE IF
ELECTRIC MOTOR
IS NOT USED
LINE SHAFT $1\frac{7}{16}$" DIA.
SPEED 200 R.P.M.
GAS ENGINE
HANGER
GRINDER COUNTERSHAFT
GRINDER
LATHE
BENCH
HANGER
LATHE COUNTERSHAFT
SOUTH BEND LATHE WORKS

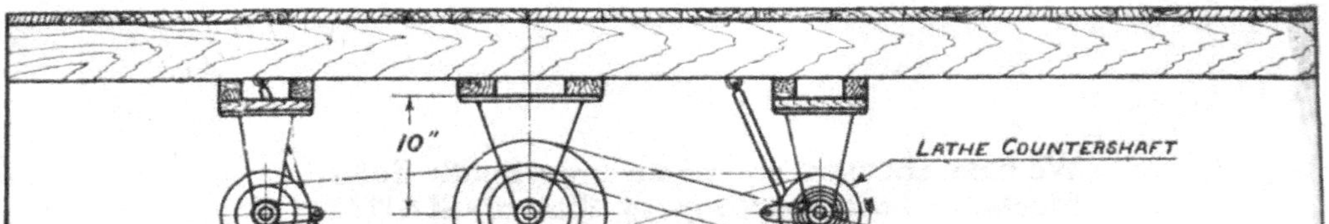
10"
LATHE COUNTERSHAFT
GRINDER
COUNTERSHAFT
LINE SHAFT $1\frac{7}{16}$" DIA.
SPEED 200 R.P.M.
ELECTRIC MOTOR
3 HORSE POWER
BENCH
LATHE
GAS ENGINE
GRINDER
SOUTH BEND

LAYOUT OF A SMALL MACHINE SHOP

Drawing on the opposite page illustrates a practical layout for a small machine shop.

The line shaft is attached to the joists about in the center of the shop and parallel to the side wall. It should be leveled perfectly so that it will run free and easy. A line shaft $1\frac{7}{16}$ inch in diameter, having a speed of 200 revolutions per minute, is recommended for the small shop. The shaft in drawing is 16 ft. long and requires three hangers, having about a 10-inch drop.

If an electric motor is used for power it should be mounted on a bracket attached to either side wall high enough so that the belt will not interfere with the workmen passing underneath.

If a gas engine is used instead of an electric motor, the engine may be located conveniently in the corner, as shown.

Countershaft for lathe should be attached to joists about 6 ft. from the line shaft and leveled carefully, the axis being parallel to that of the line shaft. The countershaft belt at the head end of lathe is straight; the reverse belt is crossed. Arrows show the direction in which the belts should run. The two pulleys on line shaft for driving countershaft are the same diameter, but some prefer the reverse belt pulley one or two inches larger in diameter, so that for thread cutting the carriage may be reversed at a high speed. Wood split pulleys, crown face, are recommended on a line shaft for driving countershaft.

A three-horse power electric motor is ample to drive the equipment shown in drawing, assuming that the lathe is not larger than a 16-inch swing.

HORSE POWER

The horse power required to drive a South Bend lathe to its full capacity under a maximum load is as follows:

11-inch lathe	....	½	horse	power
12-inch	"	½	"	"
13-inch	"	¾	"	"
15-inch	"	1	"	"
16-inch	"	2	"	"
18-inch	"	2½	"	"

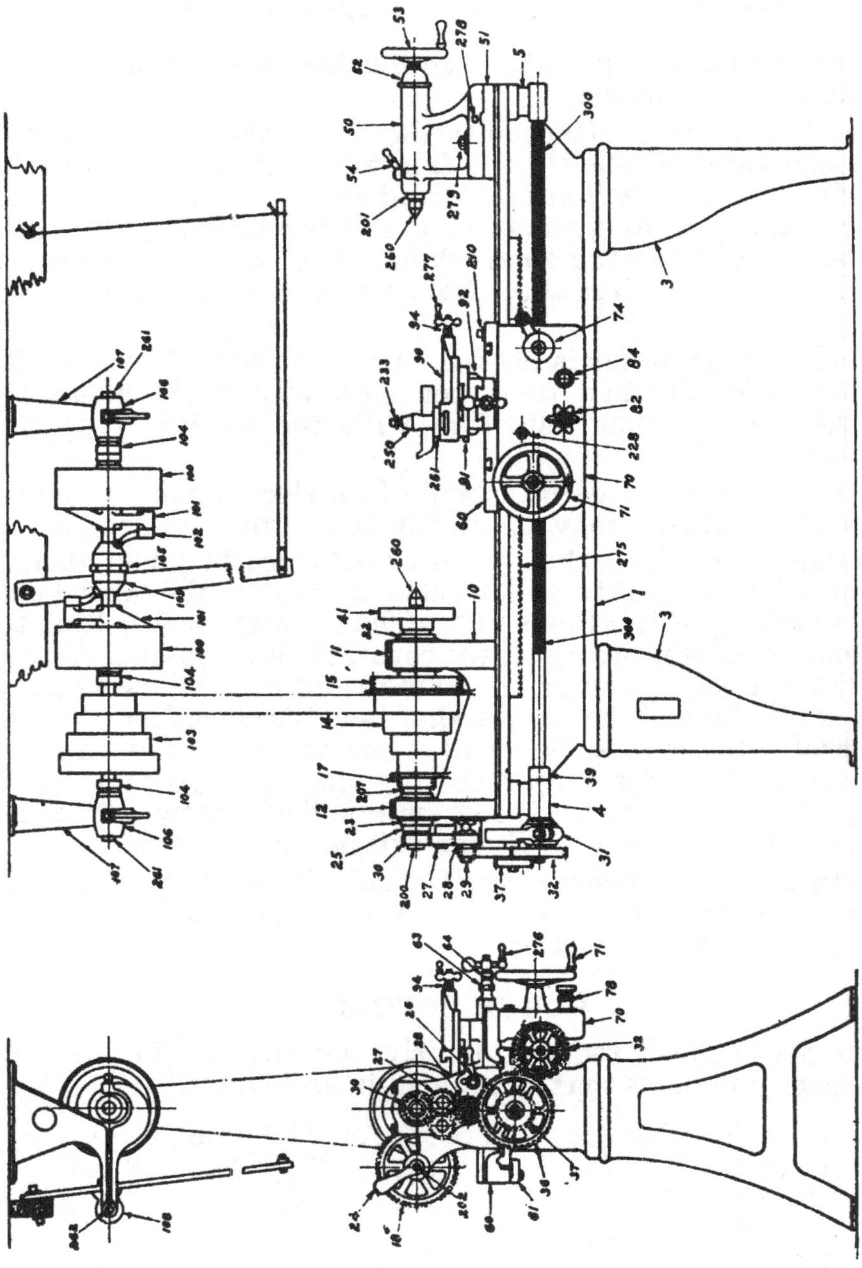

Cut of Lathe showing principal parts numbered

ORDERING REPAIR PARTS

In ordering *repair parts* for lathe, order by number of part and give *size of lathe* that the part is to be used on, because the part numbers apply to all size South Bend lathes from 11 to 18 inch swing inclusive.

NUMBER AND NAME OF LATHE PARTS ON DRAWING

NO.	
1	Bed.
3	Power Legs.
4	Lead Screw Bracket F.
5	Lead Screw Bracket R.
10	Head Stock.
11	Head Stock Cap, Large.
12	Head Stock Cap, Small.
13	Head Stock Clamp Plate.
14	Spindle Cone.
15	Bull Gear.
16	Bull Gear Clamp.
17	Cone Pinion.
18	Quill Gear.
19	Quill Sleeve.
20	Quill Sleeve Pinion.
21	Ecc. Shaft Bushing.
22	Bronze Box, Large.
23	Bronze Box, Small.
24	Back Gear Lever.
25	Spindle Take up Nut.
26	Reverse Bracket.
27	Reverse Twin Gears (2).
28	Reverse Gear.
29	Stud Gear.
30	Spindle Reverse Gear.
31	Change Gear Bracket.
32	Change Gears.
33	Change Gear Idler.
34	Bushing for Change Gear Idler.
35	Change Gear Collar on L. S.
36	Compound Idler Gear 1 to 2 Large.
37	Compound Idler Gear 1 to 2 Small.
38	Bushing for Compound Idler Gear, 1 to 2.
39	Thrust Collar on Lead Screw.
40	Large Face Plate.
41	Small Face Plate.
42	Turning Gear.
50	Tail Stock Top.
51	Tail Stock Base.
52	Tail Stock Nut.
53	Tail Stock Hand Wheel.

NO.	
54	Tail Stock Binding Lever.
55	Tail Stock Wrench.
56	Tail Stock Clamp Plate.
60	Saddle.
61	Saddle Gib.
62	Saddle Lock.
63	Cross Feed Bushing.
64	Cross Feed Gra. Collar.
65	Cross Feed Nut.
66	Plain Rest.
67	Thread Cutting Stop.
70	Apron.
71	Apron Hand Wheel.
72	Lead Screw Half Nut.
73	Lead Screw Half Nut Gib (2).
74	Nut Cam.
75	Nut Cam Washer.
76	Rack Pinion Gear.
77	Auto. Apron Worm Wheel.
78	Auto. Apron Clutch Sleeve Bushing.
79	Auto. Apron Worm Bracket.
80	Auto. Apron Clutch Sleeve.
81	Auto. Apron Clutch.
82	Auto. Apron C. F. Star Knob.
83	Auto. Apron C. F. Lever.
84	Auto. Apron C. F. Lever Knob.
85	Auto. Apron C. F. Gear.
86	Auto. Apron Idler Gear.
87	Auto. Apron Idler Gear Pinion.
90	Compound Rest Top.
91	Compound Rest Swivel.
92	Compound Rest Bottom.
93	Compound Rest End Cap.
94	Compound Rest Bushing.
95	Compound Rest Nut.
96	Compound Rest Chip Guard.

Countershaft

100	C. S. Friction Pulleys (2).
101	C. S. Friction Spiders (2).
102	C. S. Friction Fingers (2).

NO.

103 C. S. Cone.
104 C. S. Collars (4).
105 C. S. Yoke Lever.
106 C. S. Boxes (2).
107 C. S. Hangers (2).
108 C. S. Shipper Nut.
109 C. S. Yoke Cone.
200 Head Stock Spindle.
201 Tail Stock Spindle.
202 Back Gear Eccentric Shaft.
203 Apron Worm.
204 Apron Rack Pinion.
205 Spindle Sleeve.
207 Spindle Thrust Collar.
208 Apron Worm Collar.
209 Tool Post Block.
210 Carriage Lock Collar Screw.
211 Compound Rest Swivel Bolts.
212 C. G. Bracket Collar Screw.
213 Reverse Collar Screw.
214 Bull Gear Clamp Collar Screw.
215 Apron Clutch Sleeve Hex. Nut.
216 Compound Rest Swivel Stud.
217 Steady Rest Lock Bolt.
218 Auto Cross Feed Lever Stud.
219 Reverse Steel Washer ½" hole.
220 Apron Clutch Sleeve Pinion.
221 Compound Rest Bottom Gib.
222 Plain Rest Gib.
223 Auto Apron Clutch Screw.
224 Cross Feed Screw.
225 Apron Hand Wheel Pinion.
226 Tail Stock Screw.

NO.

227 Reverse Shaft or Stud.
228 Apron Rack Pinion Stud.
229 Reverse Shoulder Screws (2).
230 Compound Rest Screw.
231 Auto Cross Feed Stud.
232 Apron Half Nut Stud (2).
233 Tool Post Screw.
234 Apron Idler Gear Stud.
235 Cam Cap Screw.
238 Apron Worm Washer.
239 Comp. Rest Steel Wedge.
240 Gap Bridge Pins (2).
241 Reverse Stud Collar ¾" hole.
242 Change Gear Spindle Knob.
250 Tool Post.
251 Tool Post Ring.
252 Tool Post Wedge.
253 Tool Post Wrench.
254 Compound Rest Wrench.
257 Compound Rest Top Gib.
258 Comp. Rest C. P. Headless Set Screws.
260 Centers (2).
261 C. S. Shaft.
262 C. S. Shipper Rod.
263 C. S. Expansion Wedges.
275 Rack.
276 Cross Feed Ball Crank.
277 Compound Rest Handle.
278 Tail Stock Set Over Screws (2).
279 Tail Stock Clamping Bolt, Nut and Washer.
300 Lead Screw.

Compound Rest

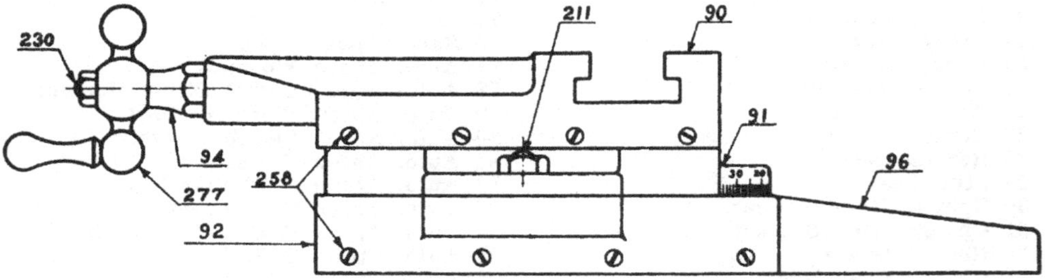

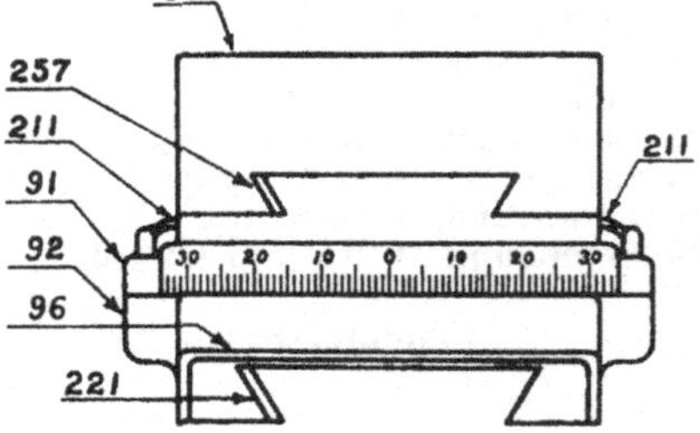

Name of Parts

NO.

90 Compound Rest Top.
91 Compound Rest Swivel.
92 Compound Rest Bottom.
94 Compound Rest Bushing.
96 Compound Rest Chip Guard.
211 Compound Rest Swivel Bolts (2).
221 Compound Rest Bottom Gib.
230 Compound Rest Screw.
257 Compound Rest Top Gib.
258 Compound Rest Cone Point Gib Screws.
277 Compound Rest Handle.

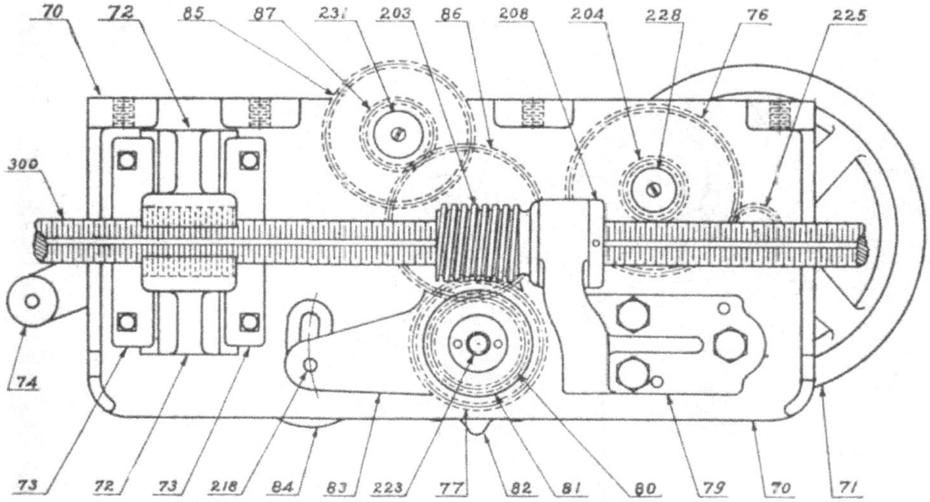

INSIDE VIEW OF AUTOMATIC APRON

Above drawing shows the inside view of automatic apron on South Bend lathes that are equipped with automatic feed. The lead screw is splined for driving worm, which operates both the automatic cross feed and the automatic longitudinal feed. This arrangement allows the thread of the lead screw to be used for screw cutting only, as the spline in the lead screw is used for driving both the automatic longitudinal feed and the automatic cross feed.

Another improved feature of this automatic apron is that the automatic cross feed and the automatic longitudinal feed can be operated only one at a time, so that when one feed is in operation it is impossible for the other feed to drop in.

Name of Parts

NO.	
70	Apron.
71	Apron Hand Wheel.
72	Lead Screw Half Nut.
73	Lead Screw Half Nut Gibs (2).
74	Nut Cam.
76	Rack Pinion Gear.
77	Auto Apron Worm Wheel.
79	Auto Apron Worm Bracket.
80	Auto Apron Clutch Sleeve.
81	Auto Apron Clutch.
82	Auto Apron C. F. Star Knob.
83	Auto Apron C. F. Lever.
84	Auto Apron C. F. Lever Knob.

NO.	
85	Auto Apron C. F. Gear.
86	Auto Apron Idler Gear.
87	Auto Apron Idler Gear Pinion.
203	Auto Apron Worm.
204	Auto Apron Rack Pinion.
208	Auto Apron Worm Collar.
218	Auto Apron C. F. Lever Stud.
223	Auto Apron Clutch Screw.
225	Auto Apron Hand Wheel Pinion.
228	Auto Apron Rack Pinion Stud.
231	Auto Apron Cross Feed Stud.
300	Lead Screw.

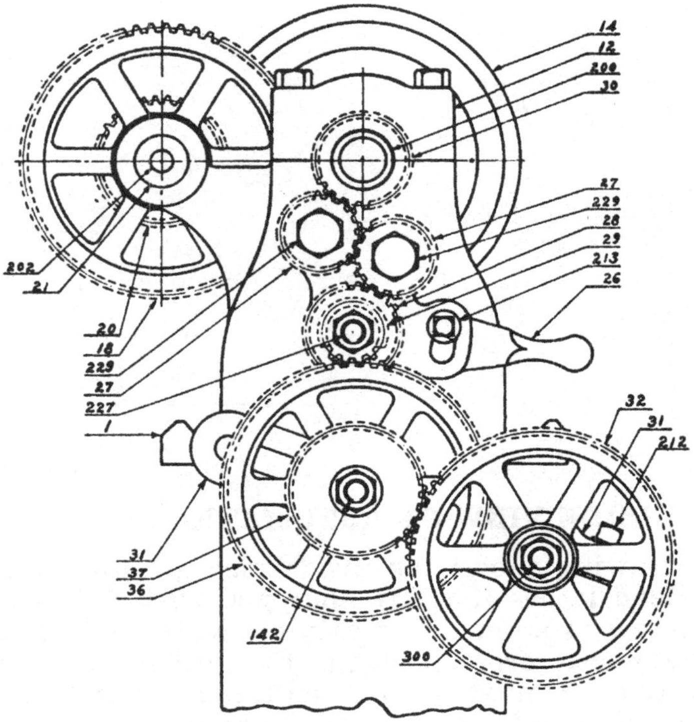

IMPROVED REVERSE

The Reverse changes the direction of the lead screw. For example, cutting threads right or left, driving the automatic feed right or left, and operating the cross feed in or out.

NO.

1 Bed.
12 Head Stock Cap, Small.
14 Spindle Cone.
18 Quill Gear.
20 Quill Sleeve Pinion.
21 Ecc. Shaft Bushing.
26 Reverse Bracket.
27 Reverse Twin Gears (2).
28 Reverse Gear.
29 Stud Gear.
30 Spindle Reverse Gear.
31 Change Gear Bracket.

NO.

32 Change Gears.
36 Compound Idler Gear, 1 to 2 Large.
37 Compound Idler Gear, 1 to 2 Small.
142 Idler Gear Bolt.
200 Head Stock Spindle.
202 Eccentric Shaft.
212 C. G. Bracket Collar Screw.
213 Reverse Collar Screw.
227 Reverse Shaft or Stud.
229 Reverse Shoulder Screws (2).
300 Lead Screw.

THE NEW LATHE

On the arrival of the new lathe remove crating and unpack the parts found in box. See that nothing is overlooked in the wrappings or excelsior. The regular equipment that is included in the price of lathe is illustrated and described on pages 58, 60 and 62. Check up the various parts with these lists.

The bright parts of the lathe are covered with grease in order to prevent rusting. Remove this grease with a rag that has been dipped in coal oil or kerosene, then wipe dry. Clean all gears thoroughly, including back gears and change gears. Clean each tooth separately, because sawdust and dirt may have lodged there while the lathe was in transit.

LOCATION OF LATHE

Select the most desirable location in your shop for the lathe. There are several conditions that will govern this spot, such as the position of line shaft, direction of the light, etc. The best results are obtained when the operator working at the lathe has the light coming from a point over his right shoulder.

Study drawings and descriptions of "Practical Layout for Small Shop," on pages 2 and 3.

The floor on which the lathe rests should be solid so that the lathe stands on a firm foundation. If there is any shake the floor should be braced from underneath. There should be a space free all around the lathe as the operator may have an occasional job where it will be necessary that he work from all sides of the machine.

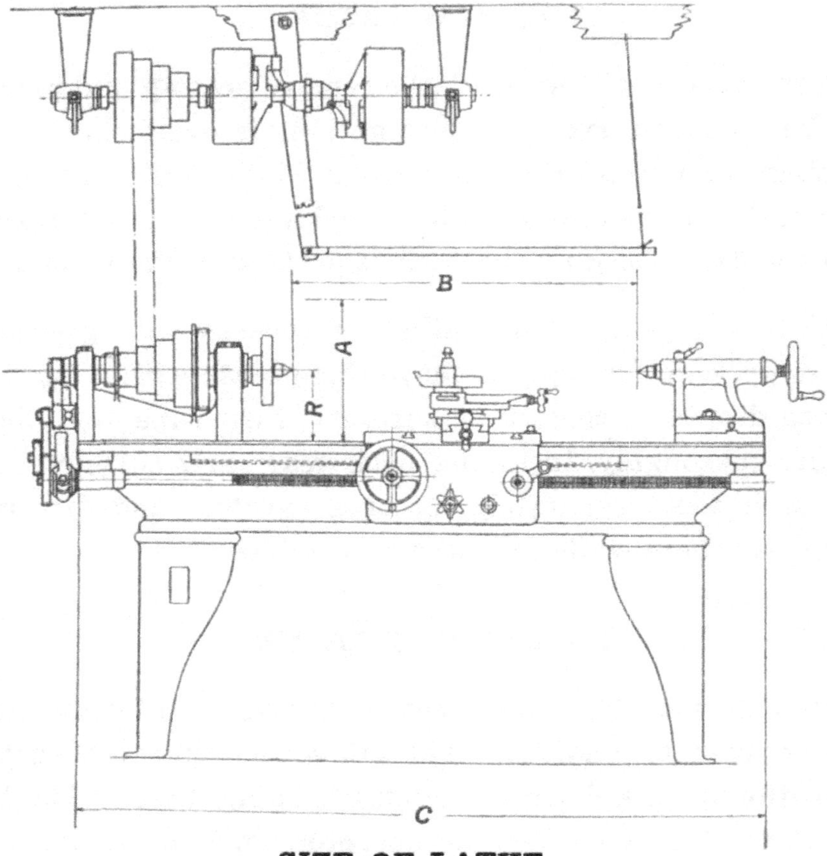

SIZE OF LATHE

The size of an Engine Lathe is determined by the *swing* and *length of bed.* See above drawing.

R represents the radius, one-half the swing.
A represents the swing.
B represents the distance between centers.
C represents the length of bed.

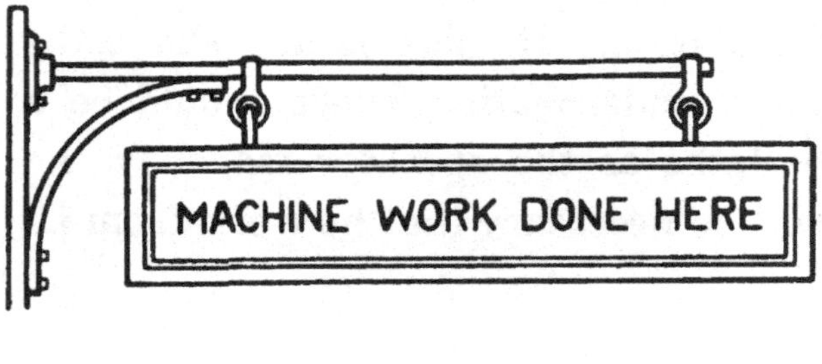

A sign that will bring business to the repair shop.

The drawing on opposite page shows the front erection view of lathe and countershaft. The height of the countershaft should be at least 5 feet from spindle in order to give proper belt distance. Seven feet high would give better results.

The countershaft may be set on either side of the line shaft being governed by the position of the lathe. See pages 2 and 3. The hangers of countershaft must be set in position so that the extension on hanger through which the shipper rod passes will point toward the line shaft, in order that the shipper rod may guide the belts.

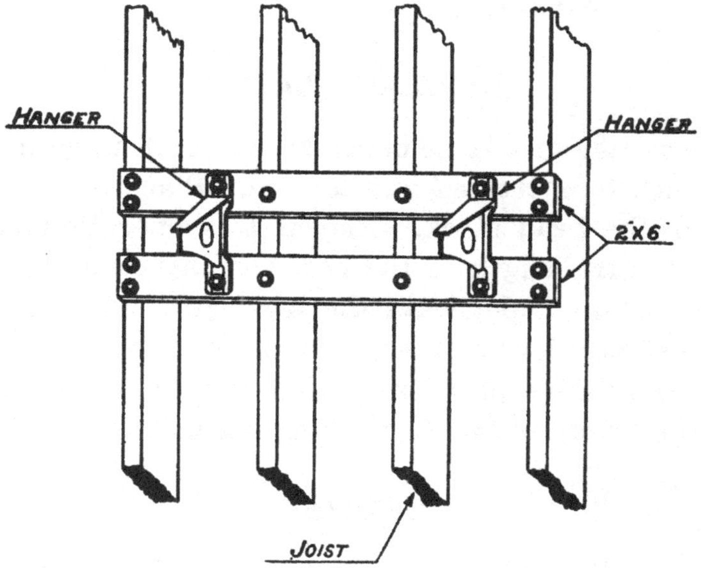

Plan of Attaching Countershaft to Joists

When the location of the lathe is settled, attach the countershaft to the ceiling. A good plan for attaching is to arrange as per drawing herewith. Assemble the countershaft complete, bolt the hangers to the 2 x 6's, in order to get the proper distance between hangers. Now remove the countershaft from the hangers, and fasten the 2 x 6's (with hangers) to the joists on ceiling. Secure with several lag screws. Then place the countershaft back in the hangers, level it up properly, adjusting so that the axis of the countershaft is parallel to that of the line shaft.

SETTING THE LATHE IN POSITION

Countershaft fastened in position, adjust the lathe so that the belt will track between spindle cone and countershaft cone. It is necessary that the axis of the lathe spindle is parallel to that of the countershaft. It is not necessary that the lathe spindle be directly underneath the countershaft, (looking at the lathe from the end) as the lathe may be from 6 to 12″ either one side or the other of a vertical line dropped from countershaft. See drawing, page 2.

LEVELING LATHE

The *accuracy* of a lathe depends a great deal upon the manner in which it is *erected* and *leveled.* Place level across the ways near the head stock. Repeat this operation at several points the entire length of the bed, leveling carefully in every direction. Shim underneath the legs with shingles until the lathe is perfectly level. When countershaft and spindle cones are in their relative position so that belts will track properly, fasten lathe firmly to the floor, using lag bolts.

BELTING

Leather belt is always recommended. In belting countershaft to line shaft two belts are used. The belt nearest the head end of lathe is usually straight, the other is the reverse or cross belt. See drawing page 2. The countershaft belts should be so arranged that when the shipper rod is thrown over in the direction of the lathe head the spindle of the lathe should revolve so that the top of the spindle cone runs toward the operator when he is in front of the lathe. A crown face pulley is used on the line shaft when its belt does not require shifting, but for a shifting belt a flat face pulley should be used.

RULES FOR CALCULATING THE SPEED AND SIZE OF PULLEYS

The driving wheel is called the driver, and the driven wheel the driven or follower.

The R. P. M. is the number of revolutions per minute.

Problem 1

The revolutions of driver and driven, and the diameter of the driven, being given, required the diameter of the driver.

RULE.—Multiply the diameter of the driven by its number of revolutions, and divide by the number of revolutions of the driver.

Problem 2

The diameter and revolutions of the driver being given, required the diameter of the driven to make a given number of revolutions in the same time.

RULE.—Multiply the diameter of the driver by its number of revolutions, and divide the product by the required number of revolutions.

Problem 3

The diameter and number of revolutions of the driver, with the diameter of the driven, being given, required the revolutions of the driven.

RULE.—Multiply the diameter of the driver by its number of revolutions, and divide by the diameter of the driven.

Problem 4

The diameter of driver and driven, and the number of revolutions of the driven, being given, required the number of revolutions of the driver.

RULE.—Multiply the diameter of the driven by its number of revolutions, and divide by the diameter of the driver.

SPEED OF LATHE COUNTERSHAFT

Size of Lathe	Size of Friction Pulley	Speed of Countershaft
11 in.	7 x 2 in.	225 R. P. M.
12, 13 in.	8 x 2½ in.	225 R. P. M.
15 in.	10 x 3 in.	200 R. P. M.
16, 18 in.	10 x 4 in.	180 R. P. M.

OIL THE LATHE EVERY DAY

Frequent oiling is necessary for a lathe or any other machine. A good grade of machine oil gives best results. After lathe is located and fastened to the floor and countershaft attached to the ceiling, the next step is to oil the revolving parts of the lathe. Every oil hole should be located and a generous supply of machine oil used flushing each revolving part, not only to give it the necessary oil, but to make sure that any dust or dirt that may have located in the bearings be washed out.

Oil the head stock spindle, also the mechanism in the apron and the lead screw bearings, all should receive special attention, so that the parts will run free and easy. The same attention should be given to the countershaft before it is attached to the ceiling and after it is set in position.

Oil the spindle cone bearings, removing two small headless set screws which will be found on the large and small steps of the cone. Give each opening a generous supply of oil, then replace the screws as they prevent dirt reaching the spindle.

Oil the back gear quill through two small oil holes for that purpose. Be sure to use plenty of oil on the lead screw and half nuts *before cutting a thread.*

Oil the head spindle bearings frequently. It is a very good plan to go over the lathe at least once a day to see that every revolving part has received sufficient oil.

Oil the countershaft twice a day for two weeks at least, until it is perfectly lubricated. Make sure to oil both loose pulleys and both boxes. If the countershaft receives this attention for the first few weeks you will avoid a great deal of trouble.

STARTING LATHE

Lathe is now ready to start. See that the shipper rod is properly adjusted. Arrange spindle cone for open belt by throwing out the back gear lever, which releases the back gears, then fasten bull gear clamp to spindle cone. Now the lathe is arranged for open belt. If the spindle runs freely on open belt stop lathe, loosen bull gear clamp, slip clamp down as far as it will go, then fasten. Throw in back gear lever connecting back gears. Lathe is now running back geared.

Never throw in back gears while lathe spindle is running.

Never throw out back gear while lathe spindle is running.

Connect reverse twin gear with spindle reverse gear and fasten the collar screw on reverse lever, locking lever in position.

Never throw in reverse gear while lathe is running.

Never throw out reverse gear while lathe is running.

Then connect gears on change gear bracket with stud gear. Now head spindle and lead screw are connected.

In this connection there should be a slight play or shake between the teeth of the two connecting gears. *This point is important.* In connecting any two gears that mesh with each other there should always be a little clearance between the top of the teeth of one gear and the bottom of the teeth of the opposite. When two or more gears are in mesh in this manner, they will operate with very little friction and give best results.

When the change gears are connected to the stud gear, lock the change gear bracket in position by fastening the collar screw in front, then start lathe, all gears being in mesh.

Never connect, or disconnect, change gear bracket gears with stud gear when lathe is running.

Oil both head end bearing and tail end bearing of lead screw.

CARRIAGE

The saddle and apron combined is called the carriage of the lathe. Oil the ways and V's of the lathe bed thoroughly, running the carriage back and forth over the oiled surface, so that the oil is distributed properly. Oil the rear saddle gib by placing some oil on your finger and spreading this oil under the rear way of the lathe where the saddle gib slides.

FACE PLATE

Before mounting a face plate on the nose of spindle all dirt should be removed from the thread of spindle and also from the threaded hole in the face plate. A few drops of oil on the spindle nose will allow the plate to screw on much easier. If the face plate screws on tightly, there must be dirt in the threads. Unscrew the plate, remove the dirt, try again. The back of the face plate should go tight up against the shoulder of spindle. The same attention should be given in attaching a lathe chuck to spindle nose.

LATHE CENTERS

The lathe centers are made of carbon tool steel. The center for the head spindle is always soft, because it revolves with the work. The center in tail spindle is always hardened, because the work revolves on this center and causes constant wear.

When one is about to do an accurate job on centers, be sure that the centers are running true. The head center should be trued up occasionally. The tail center should also be trued up when necessary. To machine a center true, remove the face plate from head spindle; see that the center is tight in the spindle, and use a flat nosed tool about 3/8 inch wide at the cutting edge. Fasten this tool in the tool post and take a light chip off the center, keeping the angle 60 degrees, as shown by center gauge in cut herewith.

Testing Angle of Center, by Center Gauge

To true up the tail center, anneal it and machine in the same manner as the head center.

Each time a center is placed in a spindle see that both the spindle hole and center are free from dirt. *Never put your finger in the spindle hole* to remove dirt while the lathe is running, but use a stick with a small piece of rag wound around it.

DIRECTION OF FEED WITH A JOB ON CENTERS

In machining a job on centers in the lathe the feed of the tool should always be, when possible, in the direction of the head spindle. The reason is obvious: When the carriage is feeding toward the head spindle and the tool taking a heavy chip, the pressure is on the head center which revolves with the work. Should the tool when taking a heavy chip feed toward the tail stock, then the pressure is on the tail center. This center is stationary, and is therefore liable to heat unless watched carefully. It is sometimes necessary to feed toward the tail stock, or from left to right, for example, in cutting a left hand thread, but whenever possible arrange the work so that you can feed from right to left, or towards the head stock.

MORSE TAPERS

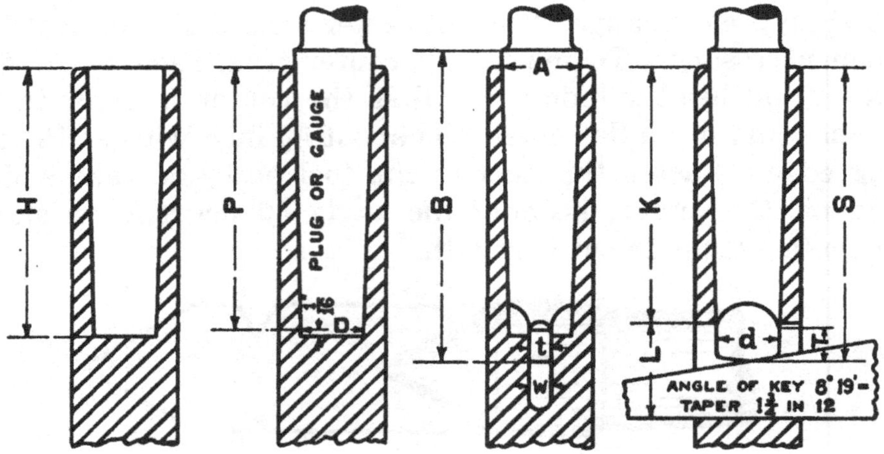

DETAIL DIMENSIONS

	NUMBER OF TAPER		0	1	2	3	4	5	6	7
	DIAM. OF PLUG AT SMALL END	D	.252	.369	.572	.778	1.020	1.475	2.116	2.750
	DIAM. AT END OF SOCKET	A	.3561	.475	.700	.938	1.231	1.748	2.494	3.270
SHANK	WHOLE LENGTH OF SHANK	B	2 11/32	2 9/16	3 1/8	3 7/8	4 7/8	6 1/8	8 9/16	11 5/8
SHANK	SHANK DEPTH	S	2 7/32	2 7/16	2 15/16	3 11/16	4 5/8	5 7/8	8 1/4	11 1/4
	DEPTH OF HOLE	H	2 1/32	2 3/16	2 5/8	3 1/4	4 1/8	5 1/4	7 3/8	10 1/8
	STANDARD PLUG DEPTH	P	2	2 1/8	2 9/16	3 3/16	4 1/16	5 3/16	7 1/4	10
TONGUE	THICKNESS OF TONGUE	t	5/32	13/64	1/4	5/16	15/32	5/8	3/4	1 1/8
TONGUE	LENGTH OF TONGUE	T	1/4	3/8	7/16	9/16	5/8	3/4	1 1/8	1 3/8
TONGUE	DIAMETER OF TONGUE	d	.235	.343	17/32	23/32	31/32	1 13/32	2	2 5/8
KEYWAY	WIDTH OF KEYWAY	W	.160	.213	.260	.322	.478	.635	.760	1.135
KEYWAY	LENGTH OF KEYWAY	L	9/16	3/4	7/8	1 3/16	1 1/4	1 1/2	1 3/4	2 5/8
	END OF SOCKET TO KEYWAY	K	1 15/16	2 1/16	2 1/2	3 1/16	3 7/8	4 15/16	7	9 1/2
	TAPER PER FOOT		.625	.600	.602	.602	.623	.630	.626	.625
	TAPER PER INCH		.05208	.05	.05016	.05016	.05191	.0525	.05216	.05208
	NUMBER OF KEY		0	1	2	3	4	5	6	7

Young man, learn the machinist's trade, learn mechanical drawing. If you master both subjects you will be a trained man and your future will be limited only by your ability.

CENTERING

To turn or machine a shaft on centers in the lathe it is necessary that the shaft be centered.

There are many ways of centering. The simplest is to chalk the end of the shaft to be centered. Scratch two lines at right angles to each other, and where these lines intersect will be the approximate center of the shaft. See Fig. 4.

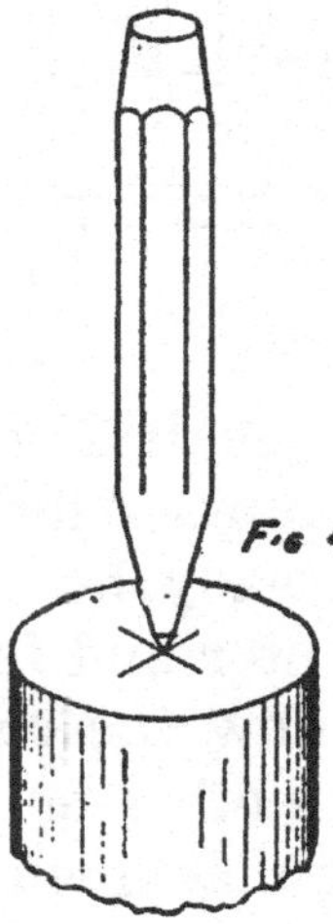
Fig 4

Drive center punch on the intersection of lines on both ends of the shaft. Place the shaft thus punched on the centers of the lathe, revolve by hand; if it does not run true, hold a piece of chalk to the shaft while revolving and it will mark the high spots. Place shaft in the vise once more, drive the center in the direction necessary to have the shaft run true.

If the piece to be machined is close to size in the rough so that very little stock is left for finishing, care should be taken to see that the bar is straight as possible and that the center holes are located accurately so that it may be true all over when finished.

Drill and countersink each end of the shaft until a depth is reached sufficient to support the shaft on the centers while it is being machined. A good method of countersinking is shown in Fig. 5, page 20.

Long shafts may be centered with a breast drill or brace and bit. The regular countersinks may be used in the same manner.

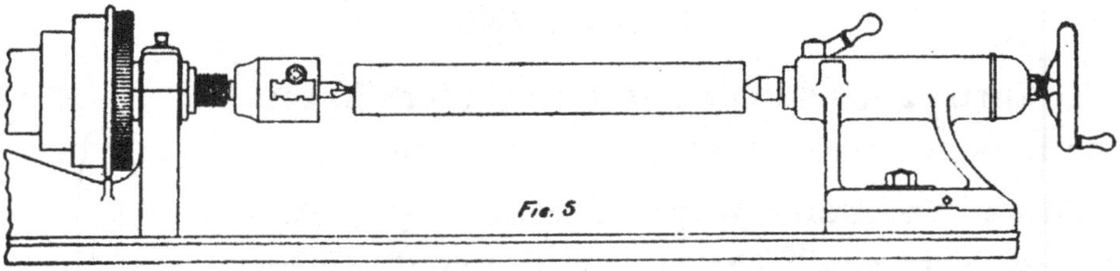

Fig. 5

COUNTERSINKING A SHAFT

A drill chuck is fitted to the head spindle of lathe holding a combined drill and countersink. (See Fig. 5-B.) The shaft has already been center punched. Place one end of shaft on tail center and feed shaft to drill by turning the wheel of tail stock. Allow the countersink to enter the proper depth, countersink the other end in the same manner. The shaft is now ready to be machined.

DRILL AND COUNTERSINK

The depth of the countersink depends upon the diameter of the shaft to be machined, and somewhat upon the depth of the cut to be taken. It is important that the countersink have the same taper as the lathe center, 60 degrees.

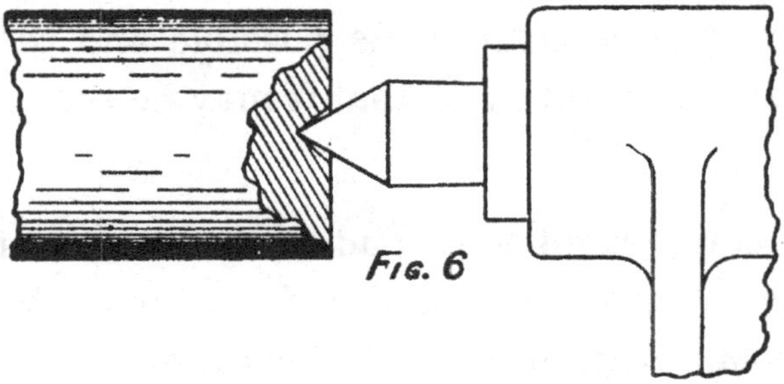

Fig. 6

IMPROPER CENTERING

Fig. 6 shows a shaft countersunk in such a way that it does not fit on the lathe center, but rests on the point only. This very soon destroys the lathe center and will also ruin the shaft.

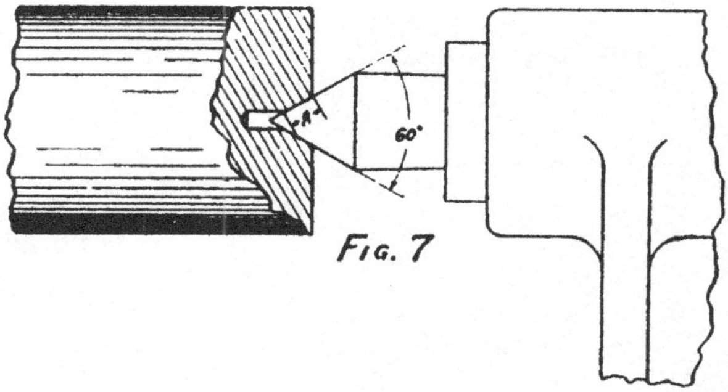

FIG. 7

PROPER COUNTERSINK

Fig. 7 shows the style of countersink which gives best results. A small hole is first drilled beyond the depths of the point of the lathe center; it is then countersunk on an angle of 60 degrees to fit the lathe center.

DRILL AND COUNTERSINK COMBINED

Fig. 5B is a combined drill and countersink, which is both the center drill and countersink. Centering may also be done by a small twist drill for the center hole and a larger twist drill ground 60 degrees following as a countersink.

Fig. 5B

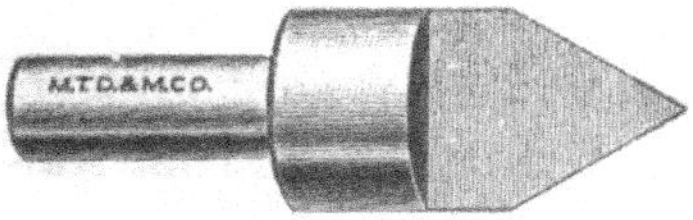

Fig. 8

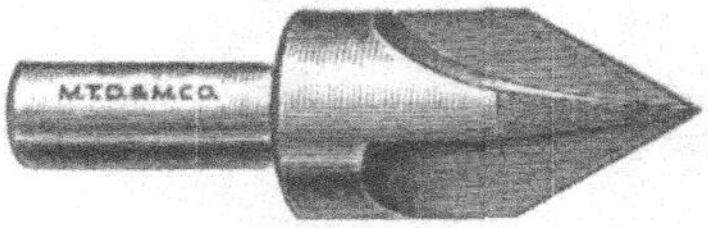

Fig. 9

Figs. 8 and 9 show two style countersink that are used to follow the small twist drill for countersinking.

TURNING A STEEL SHAFT

The illustration above shows a steel shaft being machined on a lathe between centers. The shaft is driven by a common lathe dog that is firmly attached by a set screw, the tail of the dog enters the slot in the face plate. It will be noted that a heavy chip is taken, which requires considerable power, so the lathe is running with back gears in mesh.

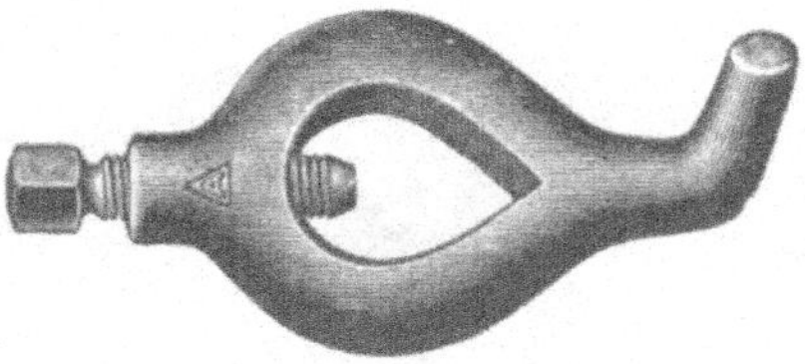

Common Lathe Dog

Clamp Lathe Dog

Henry Ford built the engine for his first car on a small lathe.

In addition to the common lathe dog, a clamp lathe dog (see illustration) is used for holding rectangular pieces that are being driven in the lathe. Sometimes a shaft is driven by holding one end in the lathe chuck, the other end being supported by a steady rest. See Fig. 18.

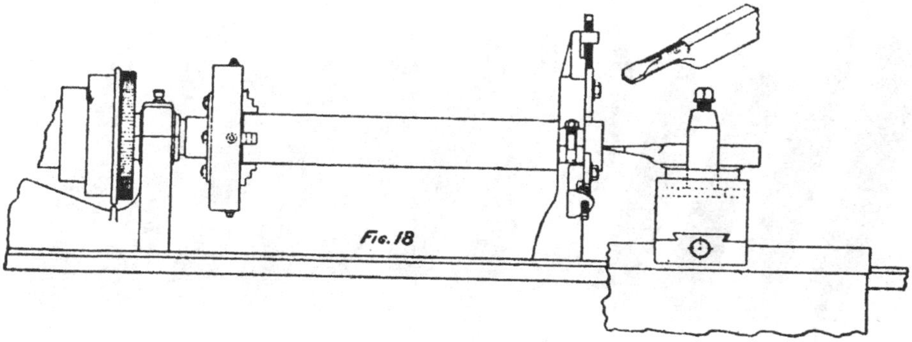
Fig. 18

A SHAFT IN THE CENTER REST

Fig. 18 shows a shaft supported by the center rest and being driven by the lathe chuck. This shaft is 3″ in diameter and 12″ long. A ¾″ hole is to be drilled its entire length, so it is necessary to center the shaft so that the drill will start true. Cut shows the centering tool held in the tool post. The point of this tool is ground exactly like the point of an ordinary flat drill. To center the shaft, start the lathe, adjust the tool in so that it is approximately at the center of the shaft. Move the tool rest, so that when the tool begins cutting it will show the exact center, and will machine a countersunk hole.

When this countersunk hole is about ⅛″ deep remove the centering tool, attach the drill to tail spindle, as shown on page 38, and proceed with the drilling, using the hand-wheel of tail stock for feeding.

FORGED STEEL LATHE TOOLS

The twelve lathe tools below have been selected as the most practical set of forged tools for general all-round work on the lathe. These tools are made both in carbon and the high speed steels. The size of the tools vary to suit the different size lathes.

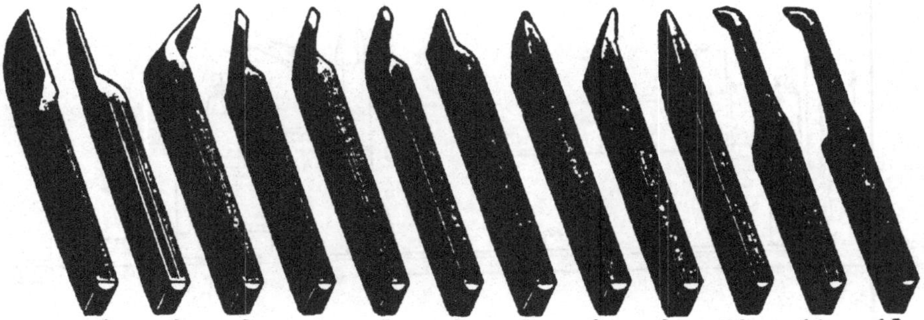

1 2 3 4 5 6 7 8 9 10 11 12

1 Left-hand Side Tool.
2 Right-hand Side Tool.
3 Right-hand Bent Tool.
4 Right-hand Diamond Point.
5 Left-hand Diamond Point.
6 Round Nose Tool.
7 Cutting-off Tool.
8 Threading Tool.
9 Bent Threading Tool.
10 Roughing Tool.
11 Boring Tool.
12 Inside Threading Tool.

LATHE TOOLS

The drawings on page 25 show the application of the above twelve lathe tools. It is not necessary that they be forged steel lathe tools. For example, the patent lathe tools shown on page 26 may be applied in the same way.

The arrow shows the direction of the feed of the tool. In some cases the tool may be fed to advantage in two directions. For example, in drawing No. 1, the arrow shows the tool feeding out toward the periphery. This tool may also be fed in toward the center. Again, in drawing No. 11, the boring tool is feeding to the left. This tool may also take a cut in feeding to the right.

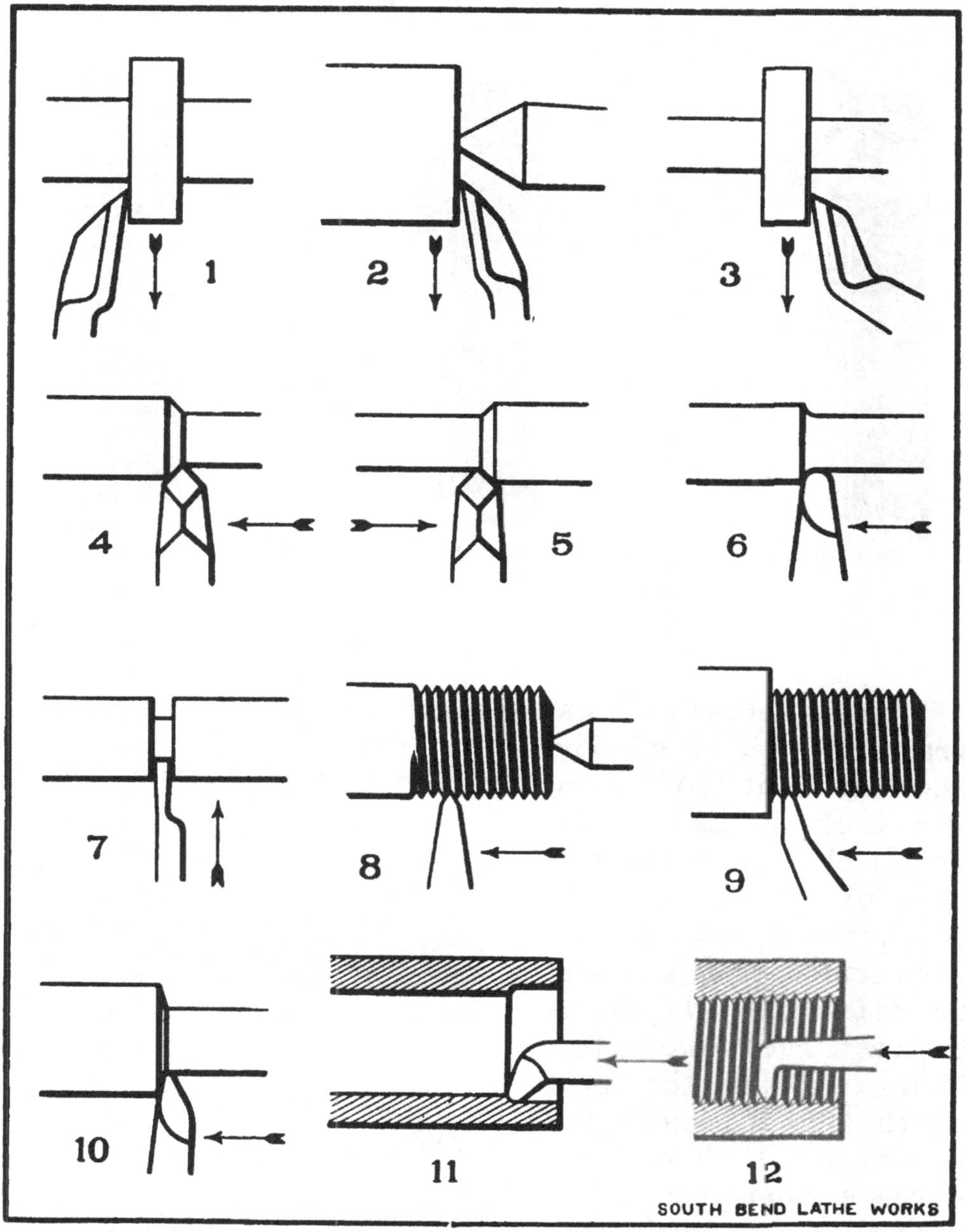

Application of Lathe Tools

PATENT LATHE TOOLS

Turning

Cut-Off

Boring

KNURLING IN THE LATHE

The cut herewith shows a piece of steel with three different grades of knurling. The knurling tool for doing this work is also shown. This tool is held the tool post of lathe; piece to be knurled is driven slowly on centers or in a chuck; tool is forced slowly into the work as it revolves, and this in turn revolves the knurl wheels, thus the impression is produced. The knurl wheels are hardened. Plenty of oil is used during the operation.

SAMPLE OF KNURLING

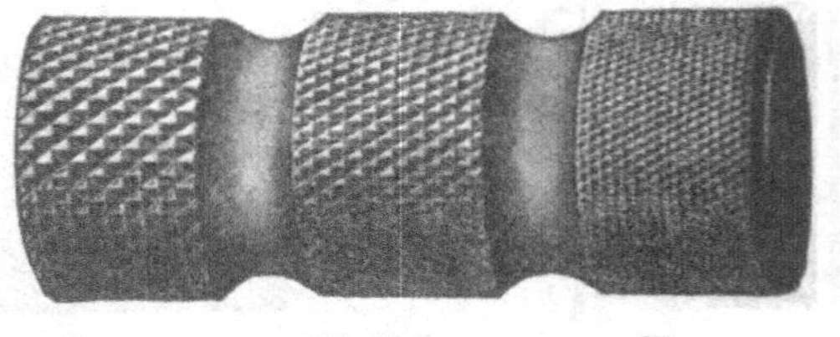

Coarse Medium Fine

Knurling Tool

POSITION OF CUTTING EDGE OF TOOL

The position of the turning tool is quite important in machining metal. In most work we find that a little above the center is the proper height. See Fig. 12.

Care should be taken that the tool does not extend too far from the tool post, especially on heavy cuts.

GRINDING THE TOOL

The efficiency of the cutting edge of a tool depends a great deal upon the way it is ground or sharpened. The tool should have plenty of clearance, a good rake and a clean cutting edge. The cutting edge of the tool is changed somewhat for various metals. After grinding a tool it would be well to dress up the cutting edge by hand with a small oil stone. This will improve its wearing qualities.

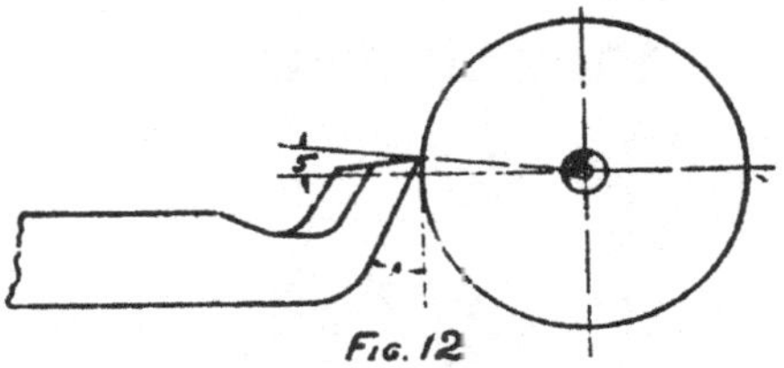

Fig. 12

FACING END OF SHAFT

When a shaft is being machined on centers, if accurate work is to be done, the end of the shaft must be faced so that it will ride on the centers evenly. A side tool is usually used to do the facing. See Fig. 11.

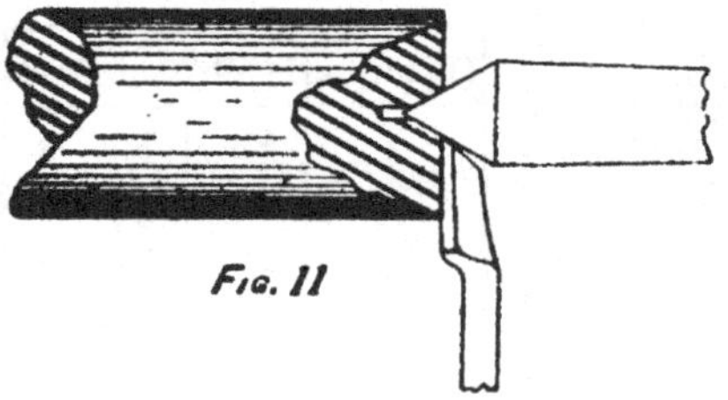

Fig. 11

In facing with a side tool it may be necessary to face into the countersink hole. On reaching the edge of the countersink tail center may be withdrawn slightly. This enables the side tool to face the end clean.

When shaft is faced ready for general machining, place a drop of oil on the tail stock center, also a drop in the countersunk hole of shaft. Adjust the tail stock center so that the shaft has a slight play on the centers (it must not turn hard on the centers), then fasten the tail stock spindle by the binding clamp. *Keep oil on the tail center.*

STANDARD SCREW THREADS

There are several different standards for the various screw threads, but general practice favors the U. S. Standard.

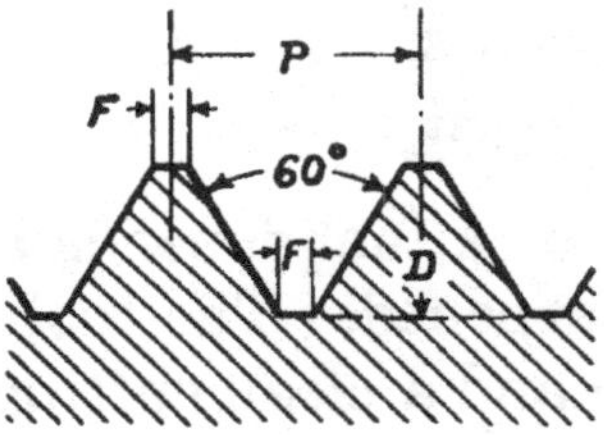

FORMULA

$$P = \text{PITCH} = \frac{1}{\text{No. THREADS PER INCH}}$$

$$D = \text{DEPTH} = P \times .64952$$

$$F = \text{FLAT} = \frac{P}{8}$$

UNITED STATES STANDARD SCREW THREADS

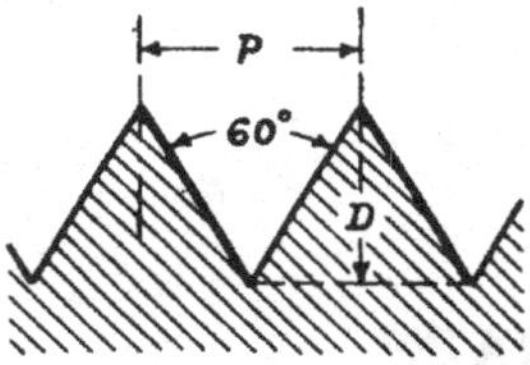

FORMULA

$$P = \text{PITCH} = \frac{1}{\text{No. THREADS PER INCH}}$$

$$D = \text{DEPTH} = P \times .86603$$

SHARP V STANDARD SCREW THREADS

TABLE SHOWING PITCH OF THREAD and Sizes of Tap Drills for Standard Thread

Diam. Tap in Ins.	Thds. per inch	Size of Drill Ins.	Diam. Tap in. Ins.	Thds per Inch	Size of Drill Ins.	Diam. Tap in Ins.	Thds. per Inch	Size of Drill Ins.
1/4	20	3/16 in.	13/16	11	37/64	1 1/4	7	1 5/64
5/16	18	1/4 in.	3/4	10	5/8	1 3/8	6	1 11/64
3/8	16	5/16 in.	11/16	10	11/16	1 1/2	6	1 19/64
7/16	14	3/8 in.	7/8	9	47/64			
1/2	13	11/32 in.	15/16	9	51/64			
1/2	12	13/32 in.	1	8	27/32			
9/16	12	29/64 in.	1 1/8	7	61/64			
5/8	11	33/64 in.						

A. L. A. M. Standard Thread and Drill Sizes for Taps

Diam. in.	Pitch	Tap Drill
1/4	28	No. 4
5/16	24	17/64 in.
3/8	24	21/64
7/16	20	3/8
1/2	20	7/16 in.
9/16	18	1/2
5/8	18	9/16
11/16	16	39/64
3/4	16	43/64
7/8	14,18	51/64, 13/16
1	14	59/64

MEASURING SCREW THREADS

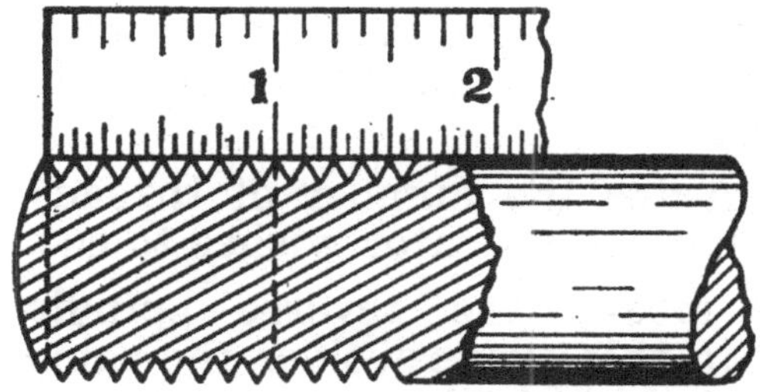

FIG. 70

To find the pitch of a screw when a thread gauge is not convenient, place a scale on the screw (Fig. 70) so that the end of the scale is opposite the top point of any thread; count the number of spaces under the scale between the threads, for a distance of one inch, viz: There are eight spaces underneath the scale in one inch, therefore, the screw is 1/8" pitch or eight threads per inch.

Another method is to place the scale as shown in Fig. 70, and count the top of the threads for a distance of one inch, omitting one thread. The reason for omitting one thread may be seen by following the two dotted lines drawn from the top point of the first and 9th threads; count the number of complete threads between the dotted lines at the bottom of the screw, and you will find it to be eight.

THREAD CUTTING

In *cutting screw threads* on the lathe the carriage is *always driven* by clamping the half nuts on the *lead screw*, therefore, the *friction feed should never be used when thread cutting.* See that the feed knob is not tight, but loose or neutral so that neither the power cross feed nor the longitudinal feed can possibly get into action while cutting threads.

Pitch and lead do not always mean the same thing. The pitch of a thread is the distance from the center of one thread to the center of the next. The lead of a screw is the distance the nut will advance in one revolution of the screw. Where a screw is single thread the pitch and lead are the same. But when a screw has a double, triple, or multiple thread, the lead is as many times the pitch as there are multiple threads in the screw.

CHANGE GEARS FOR THREAD CUTTING

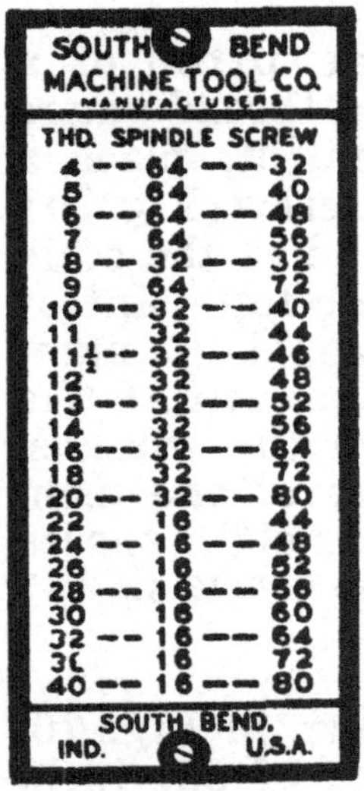

THD.	SPINDLE	SCREW
4	64	32
5	64	40
6	64	48
7	64	56
8	32	32
9	64	72
10	32	40
11	32	44
11½	32	46
12	32	48
13	32	52
14	32	56
16	32	64
18	32	72
20	32	80
22	16	44
24	16	48
26	16	52
28	16	56
30	16	60
32	16	64
36	16	72
40	16	80

Index Plate

An index plate is always attached to a South Bend lathe, showing the necessary gears to cut the desired thread. The headings of the three columns shown in plate are as follows:

Thread, means the number of threads to the inch that the operator wishes to cut.

Spindle, means the reverse spindle stud, which in this case revolves the same speed as the lathe spindle.

Screw, means lead screw, which drives the carriage of the lathe.

THREAD CUTTING

To cut a 16 thread on a No. 34 South Bend lathe, the index plate in cut shows that we should have a 32-tooth gear on the spindle and a 64-tooth gear on the screw. In order to connect these two gears we use what is called an Idler gear. The large gear of the 2 for 1 Compound Gear which is furnished with the machine will be found the most convenient Idler for this purpose. (See drawing, Fig. 13.)

Drawing, Fig. 13, shows the arrangement of gears for cutting the 16 threads as per example above. This arrangement is called "simple gearing."

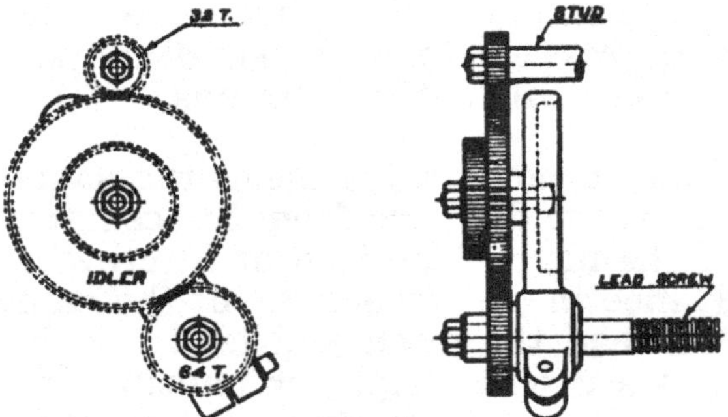

FIG. 13

Simple Gearing

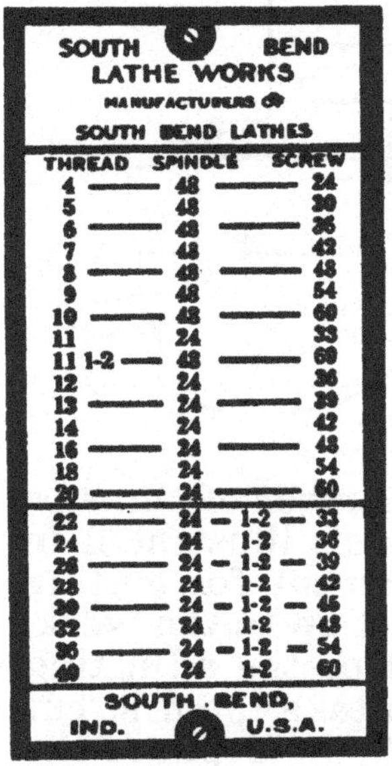

SOUTH BEND
LATHE WORKS
MANUFACTURERS OF
SOUTH BEND LATHES

THREAD	SPINDLE		SCREW
4	48		24
5	48		30
6	48		36
7	48		42
8	48		48
9	48		54
10	48		60
11	24		33
11 1-2	48		69
12	24		36
13	24		39
14	24		42
16	24		48
18	24		54
20	24		60
22	24	1-2	33
24	24	1-2	36
26	24	1-2	39
28	24	1-2	42
30	24	1-2	45
32	24	1-2	48
36	24	1-2	54
40	24	1-2	60

SOUTH BEND,
IND. U.S.A.

Index Plate

INDEX PLATE

The index plate shown in cut herewith is attached to the eighteen-inch South Bend lathe. It will be noted that when cutting 22 thread, or greater, that a Compound Idler gear 1-2 is used.

COMPOUND GEARING

The compounding of gears for thread cutting is necessary when a pitch of the thread to be cut requires gears with a great number of teeth, and to avoid having large gears or a great number of small gears we resort to compounding.

COMPOUND GEARING

Drawing, Fig. 14, shows the arrangement of compound gear for cutting a 32 thread on an 18-inch South Bend Lathe.

Index plate above shows that for cutting a 32 thread there should be a 24-tooth gear on the spindle and a 48-tooth gear on the screw.

Index plates also calls for 1 to 2 gear, so we use the regular 1 to 2 compound gears No. 36 and No. 37 furnished with the machine for this purpose. It will be necessary to attach an Idler gear in this case to connect the smaller of the compound gears with the screw gear, so we furnish an Idler gear for this purpose No. 33.

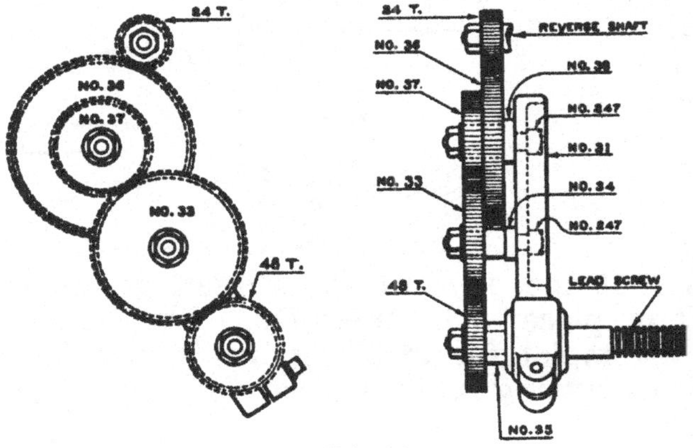

FIG. 14

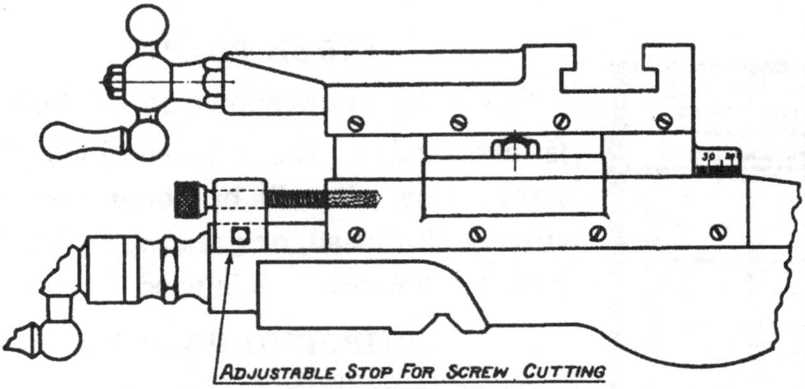

Thread Cutting Stop

In cutting a thread it is necessary that a number of chips be taken. An adjustable stop for screw cutting, as illustrated in drawing above, is a very convenient arrangement for getting the depth of the chip. After bringing the tool up to the work, fasten the stop in position so that the head of the adjusting screw is close up to the stop. When ready to take the first cut turn this screw about ¼ of a turn. This limits the depth of the cut. Repeat this operation on each cut until the screw is finished.

SETTING OF THREAD TOOL

The cutting edge of thread tool should be ground to an angle of 60 degrees. The height of the cutting edge of the tool should be even with the lathe center. A thread or center gauge is used for the setting of the thread tool in order to give the proper angle on the thread. See Fig. 15.

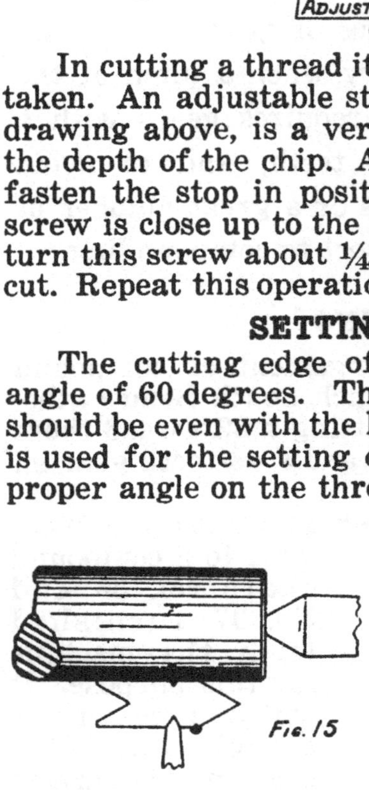

Fig. 15

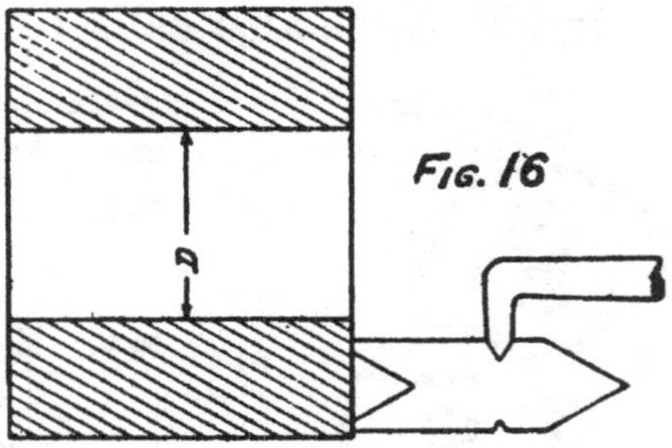

Fig. 16

Adjust the tool so that the edge of the thread gauge is parallel with the shaft or piece to be threaded. This thread gauge is also used as a templet for grinding the thread tool. Setting the thread tool for cutting internal threads is shown in Fig. 16.

THE FIRST CHIP (THREAD CUTTING)

The thread tool being set and the change gears arranged to give the desired pitch we are now ready for cutting the thread. If the material to be machined is mild steel plenty of oil should be used on the tool. The adjustable stop will regulate the depth of cut. The beginner should take very light chips. A little experience will teach him the proper depth.

Before taking the first chip see that the dog on the shaft to be threaded is tight. See that there is oil on the tail center where it enters the shaft. See that your thread tool is fastened firmly in the tool post and that the automatic feed in the apron is not in action. See that the half nuts in apron are oiled thoroughly and that there is also oil on the lead screw.

Bring the point of the tool to the end of the shaft, clamp the half nuts firmly on the lead screw, and start the lathe. For the first cut let the tool enter only far enough to make a slight scratch as far as you wish the thread to extend. When you reach the end of the cut draw the tool out by turning the cross feed screw. Reverse the shipper rod. This reverses the direction of the lathe spindle and allows the carriage to travel back automatically. Usually two complete turns of the cross feed screw will withdraw the thread tool far enough to clear the thread, so that the tool will not destroy the thread on the return of the carriage. Continue this operation until thread is finished.

Do not open the split nut to reverse carriage until you have finished cutting the thread. (See page 34, fourth paragraph.)

Piece being threaded should revolve slowly. On soft machine steel we would recommend a cutting speed of 25 feet per minute for the beginner. For example, cutting an 8-thread on 1½-inch soft steel shaft. The belt should be on the second slowest speed with back gear in.

GRINDING TOOL AFTER THREAD HAS BEEN STARTED

If it is necessary to remove the tool for grinding before thread cutting is finished, take the tool out and grind it, adjust the thread tool as before, setting it opposite the thread groove. Turn the spindle forward by hand by pulling on the belt, and again test to see if the point of the tool is exactly opposite the thread in the work. If it is not opposite drop the reverse gear, disconnecting the lead screw, and turn the spindle forward by hand until the tool is exactly opposite the groove. Then connect the reverse gear as before. This throws in the feed, and you may proceed with the cutting.

In turning the spindle by hand *always turn it forward.* If you turn it backward there will be a back lash and it will not show the true position of the tool.

A lathe is even geared when the revolutions of the spindle and the revolutions of the spindle stud are the same.

On a lathe that is even geared, when cutting a screw having the number of its threads per inch exactly divisible by the number of threads per inch in the lead screw, it is not necessary to reverse,—the lathe may run in one direction all the time. When the tool has fed to the end of the cut draw it out as before, open the split nut, and reverse the carriage by hand. Throw in the split nut again, take another chip, and repeat this operation until the screw is finished. For example, if your lead screw is 8-thread and the screw that you wish to cut is divisible by eight, such as 8, 16, 24, 32, 40, etc., the above rule may be followed.

Never remove the driving dog from the screw that you are cutting until the thread is finished. Always put the tail of the dog into the same slot of face plate after testing the screw for size.

When cutting threads on soft steel or iron use plenty of oil on the point of the cutting tool. If one has many threads to cut lard oil is recommended.

In cutting threads on cast iron, brass, aluminum, etc., oil is never used. The exception is, when threads are being cut by a die.

TURNING TAPER

In straight turning on the lathe the tail stock top and base are clamped at zero, as shown in Fig. 21. For taper turning the tail stock top is set off center from its base to get the desired angle of the taper.

To set over the tail stock of lathe for turning taper, loosen the tail stock clamp and unscrew set screw "F" the distance you think necessary to get the desired taper. Screw in the set screw "G" a similar distance, until it strikes the tail stock hub. Then clamp the tail stock to bed and take a chip. Test the angle of taper on the turned piece to see if it is correct.

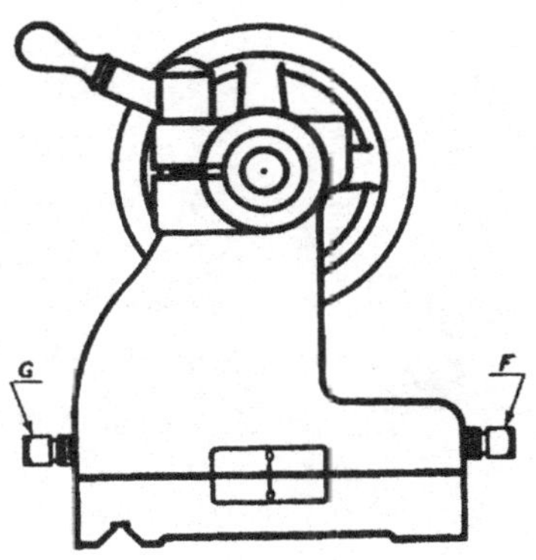

FIG. 21

In testing the taper of a piece, insert it in the taper hole, which it is to fit. Carefully shake with thumb and finger. If there is a shake it will show that the taper is too great or not great enough; adjust the tail stock accordingly, take a light chip and make another test. For a final test, make a chalk mark along the element or side of taper of the piece. Place the tapered piece in the hole and turn carefully by hand. If there is a high spot, or if the taper is not correct, it will show on the chalk mark. If it is a perfect fit it will indicate along the entire length of the taper.

Fig. 22 illustrates an automobile axle being turned taper on a 16-inch South Bend lathe.

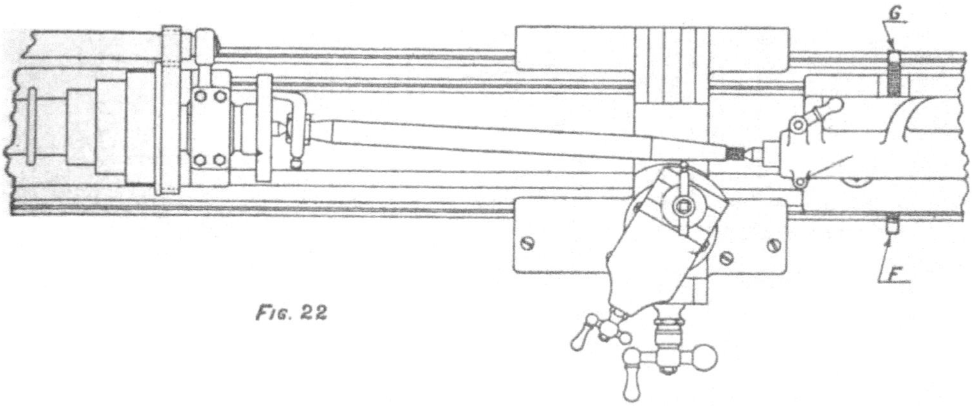

FIG. 22

Turning Taper on an Automobile Axle

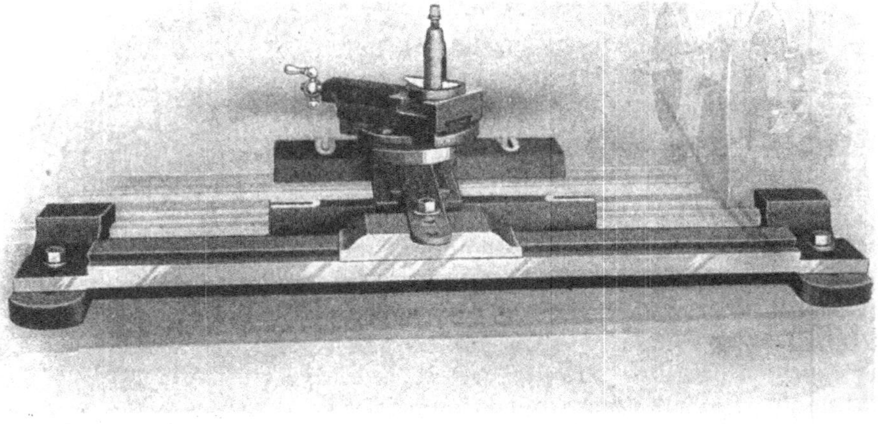

TAPER ATTACHMENT FOR SOUTH BEND LATHES

The illustration shows an efficient taper attachment for lathe. This attachment is fitted to the rear V of the lathe by two clamps. These clamps may be adjusted along the entire length of the bed, so that the operator may turn taper in any position the whole length of the lathe.

TRUING A VALVE

Fig. 590 shows a practical, and the only correct way of facing a valve. This is the method used by the valve manufacturer. The stem of the valve is held in a three-jaw Universal Chuck so that it runs exactly true. The head of the valve, which is centered, is held on the tail center of lathe. Adjust the graduated compound rest to the correct angle of the valve, and with a round nosed tool in the tool post, feed forward, taking one or two chips as required. This operation shows the advantage of having a lathe equipped with a compound rest.

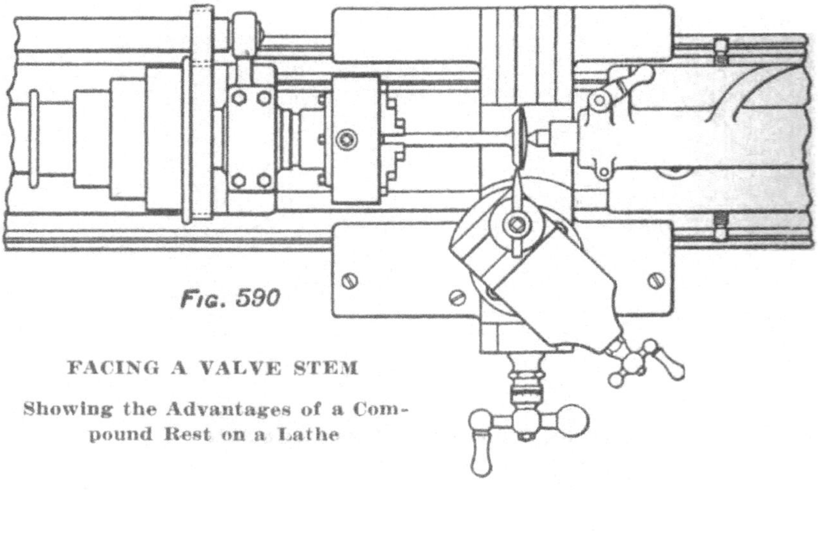

FIG. 590

FACING A VALVE STEM

Showing the Advantages of a Compound Rest on a Lathe

GRINDING ATTACHMENTS FOR LATHE

The illustration shows the practical application of a grinding attachment fitted to the lathe. The operation shows a round blanking die being ground for clearance. The grinding attachment is held in the tool post, the emery wheel being driven by an overhead drum that is usually about 10 to 14 inches in diameter.

Grinding a Blanking Die on Lathe

SUGGESTIONS ON EMERY WHEEL

Keep emery wheels true. Frequent truing will prolong the life of wheels as well as increase production, and reduce danger. Dressing is not truing, but sharpening the wheel.

Keep all rests adjusted close to the wheel, otherwise work is liable to be caught and injury result.

TABLE OF GRINDING WHEEL SPEEDS

Diam. Wheel	R.P.M. for Surface Speed of 4,000 ft.	R.P.M. for Surface Speed of 5,000 ft.
1 inch	15,279	19,099
2 "	7,639	9,549
3 "	5,093	6,366
4 "	3,820	4,775
5 "	3,056	3,820
6 "	2,546	3,183
7 "	2,183	2,728
8 "	1,910	2,387
10 "	1,528	1,910
12 "	1,273	1,592
14 "	1,091	1,364

Wheels are run in actual practice from 4,000 to 6,000 feet surface speed per minute.

DRILLING AND FACING ON THE ENGINE LATHE

The illustration above shows a number of steel discs that have been drilled and faced in one chucking on a No. 40 16-inch South Bend lathe. A Universal chuck is fitted to the spindle nose, and a drill chuck fitted to the tail spindle of the lathe.

A lathe can be rigged up with a few simple attachments to turn out a great many jobs at a productive cost equal to a high-priced special type machine, while the cost of the lathe is perhaps only one-fourth of that of the special machine, and when the job is finished you have your engine lathe for use on general machine work.

George Westinghouse got his early mechanical experience on a small screw cutting engine lathe.

USING THE LATHE AS A DRILL PRESS

The illustration shows a drilling job being done on the lathe. The drill is 1 inch diameter, the piece to be drilled is steel 1 inch thick. The cutting feed is operated by the hand wheel of tail stock. Drilling of almost any nature can be done in a lathe to advantage. The size of hole may range all the way from $\frac{1}{16}''$ to $1\frac{1}{2}''$ in diameter.

In the drilling operation above the back gears are in mesh on the lathe, so that the power delivered at the point of the drill is equal to the power of a 24″ back geared drill press. Practically any drilling job that can be done on the drill press may also be done on the lathe, because the lathe is a universal machine for the general machine shop.

Fitted to a No. 34—13" South Bend Lathe

SOUTH BEND MILLING AND KEY-WAY CUTTING ATTACHMENT FOR LATHES

The South Bend attachment illustrated above is practical in the shop because it equips the lathe for doing a great deal of work that otherwise could be done only on the shaper or milling machine.

The depth of the cut is controlled by the feed of the carriage, the length of the cut by the cross feed screw, and the graduated screw at the top takes care of the vertical motion. The attachment swivels all the way around like the compound rest, and is graduated in degrees. In addition it swivels on the upright angle plate 180 degrees, and is graduated. There is a graduated collar on the vertical screw reading in one-thousandths of an inch.

Price No. 3 Attachment, $35.00.

SQUARING A STEEL SHAFT IN THE LATHE

Illustration Fig. E shows a No. 3 attachment fitted to a lathe squaring a 1½" steel shaft. A spiral end mill is fitted into the taper of the spindle. The shaft is fed horizontally across the face of the end mill to the desired depth. Then by using the vertical feed you can get a clear sharp corner.

An end mill cutting in the above manner does not need near as much power as if it were cutting on the face, and it makes a much cleaner job.

Fig. E

KEY SEATING WOODRUFF SYSTEM

In Fig. D the illustration is taken from the back of lathe, and shows the attachment holding an inch and a half steel shaft which has been key-seated for the Woodruff system of key. A special chuck is fitted to the spindle lathe for holding the cutter.

Fig. D

KEYSEATING A STEEL SHAFT

Fig. B shows No. 3 attachment cutting a key-way ½ inch wide and 3-16 inch deep in a 2-inch steel shaft. The work is being done on a 13-inch lathe.

Fig. B

Milling Cutter

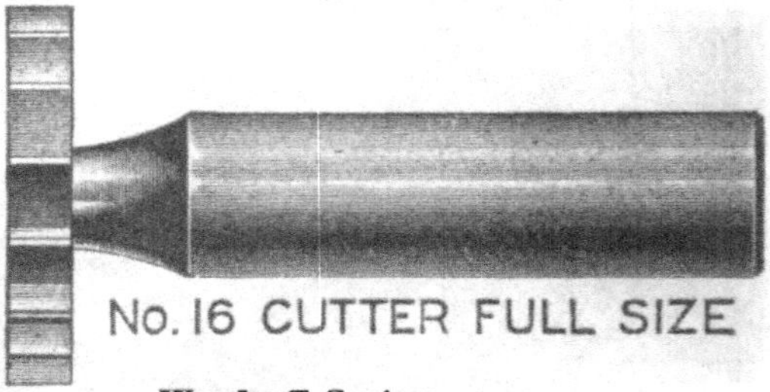

Woodruff System

The above illustration shows a Key Seat Cutter for Woodruff system of keying. In ordering a key seat cutter of this kind, give the diameter and the width of face of the cutter. Prices of any size cutter on application.

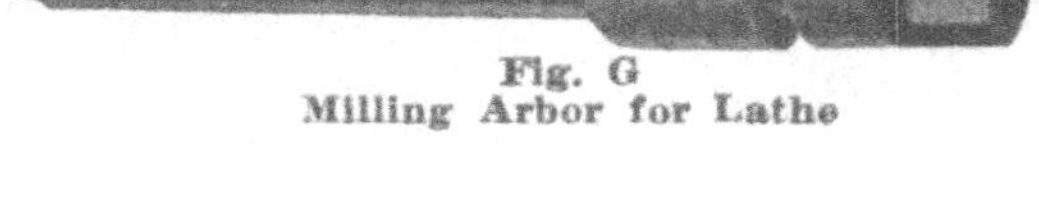

Fig. G
Milling Arbor for Lathe

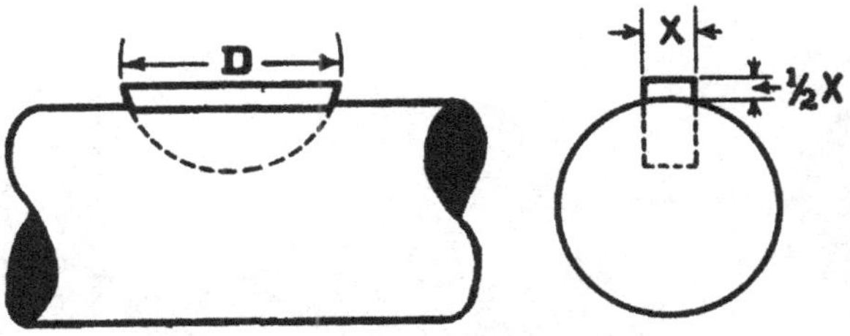

Woodruff Key-Way

The drawing herewith shows a shaft that has been milled for a Woodruff key, with key inserted. "X" equals the thickness of key. The key should project above the shaft one-half its thickness.

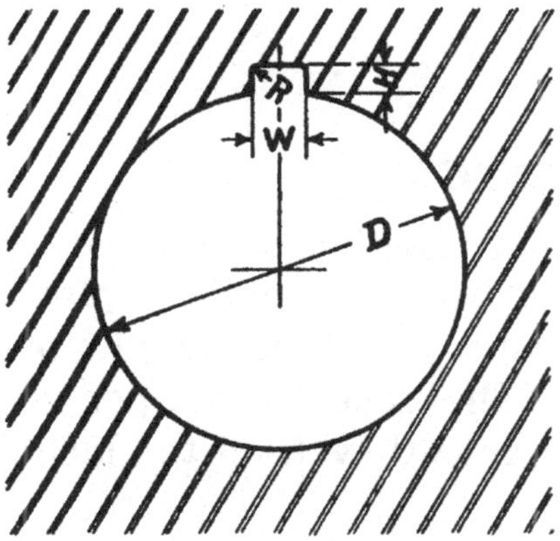

STANDARD KEY-WAYS FOR PULLEYS AND SHAFTS

This diagram shows the recognized standard for the depth and width of key-way in pulleys. The same formula of course may be used for the depth and width of key-way in shaft.

Here we give a list of the standard sizes of key-ways both for pulleys and shaft.

Diameter (D) of Hole	Width (W) of Keyway	Depth (H) of Keyway	Radius (R)
3-8" to 9-16"	3-32"	3-64"	
5-8 to 7-8	1-8	1-16	.030
15-16 to 1 1-8	5-32	5-64	.035
1 3-16 to 1 3-8	3-16	3-32	.040
1 7-16 to 1 3-4*	1-4	1-8	.050
1 13-16 to 2*	5-16	5-32	.060
2 1-16 to 2 1-2	3-8	3-16	.060
2 9-16 to 3	7-16	3-16	.060

A man that is a good lathe hand can run any kind of a machine tool.

BORING IN THE LATHE

The above cut shows a transmission case being bored in a lathe. The tool rest has been removed and an auxiliary plate bolted to the saddle. This plate may be adjusted for height by the collars or washers underneath. The case is clamped in position on the plate, and the boring bar is driven on the lathe centers.

The drawing on the opposite page shows the construction of different boring bars:

Fig. 1—Shows a boring bar equipped with a double edge cutter held by a key. Boring bar is slotted as in Fig. 2. Fig. 3 shows an end view of the double edge cutter in operation.

Fig. 4—Shows a boring bar with a fly cutter, which may be adjusted.

Fig. 6—Shows the construction of a boring bar for boring holes of large diameter. A cast iron disc is keyed to the bar, an adjustable fly cutter is set in this disc and fastened by a set screw. If one wished to bore into a sharp corner to the bottom of a hole, set a fly cutter in the disc so that it will project from the corner on an angle. In this manner one can finish up sharp corners in a cylinder.

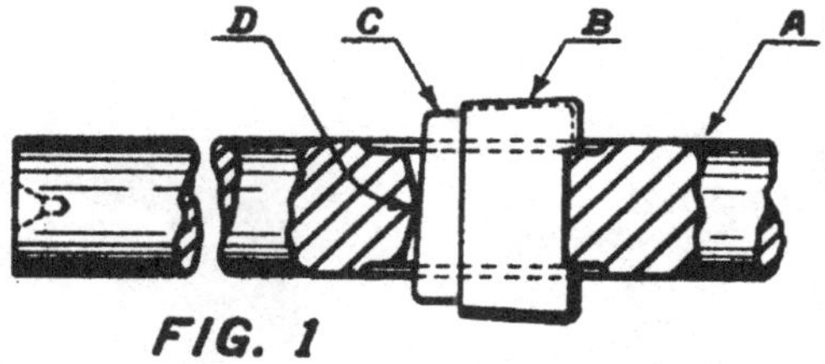

FIG. 1

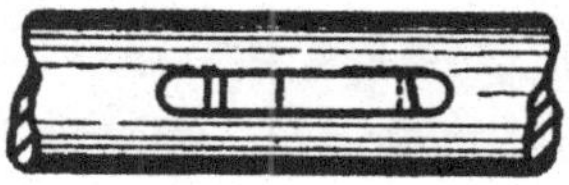

FIG. 2

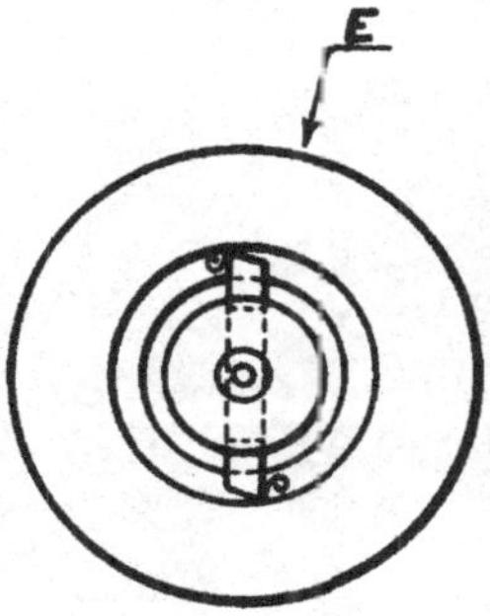

FIG. 3

FIG. 4

FIG. 5

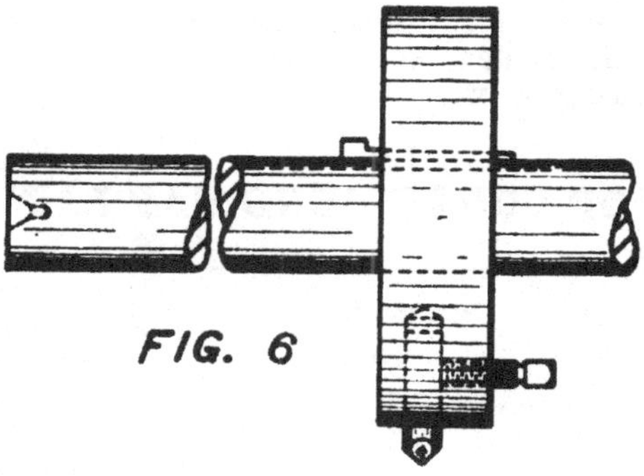

FIG. 6

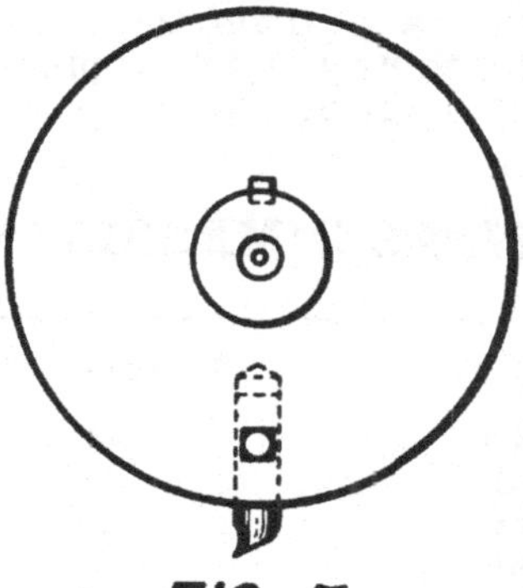

FIG. 7

SOUTH BEND LATHE WORKS

Construction of Boring Bars

16-INCH LATHE BORING A 30-INCH FLY WHEEL

Gap Bed Lathe Equipped with Raising Blocks

The practical application of the gap bed and raising blocks to the lathe may be seen in the above where the operator was able to do a job on a 16-inch lathe that otherwise would require a 30-inch lathe.

The gap bed lathe with bridge is a very practical tool for the general machine and repair shop that is called upon to do a great variety of work. If a job of large diameter comes in it is a simple matter to remove the bridge, machine the job, replace the bridge, so the lathe may be used as regular straight bed lathe for ordinary work.

PRINCIPAL DIMENSIONS OF SOUTH BEND GAP LATHES

	Swing over Gap	Width of Gap	Price extra for Gap and Bridge
11″ Lathe	15″	5″	$17.00
12″ Lathe	17″	6″	18.00
13″ Lathe	19″	7″	22.00
15″ Lathe	22″	8″	26.00
16″ Lathe	24″	8¾″	30.00
18″ Lathe	26″	10″	40.00

Note: Raising blocks may be attached to Gap Bed Lathes at the same price as that of Straight Bed Lathes.

RAISING BLOCKS

A 15-inch Lathe Blocked to Swing 22 Inches

The above cut illustrates a 15-inch lathe equipped with raising blocks to swing 22 inches over the bed. The raising block equipment includes blocks for head stock, tail stock, tool rest, center rest, also the necessary bolts and screws for attaching blocks to lathe.

The advantage of Raising Blocks on a lathe is that they increase the swing the entire distance between centers and is the most practical and economical method of increasing the swing of a lathe.

Price of Raising Blocks for the Different Size Lathes as Follows:

	Price
11″ Lathe Blocks to swing 14″ for turning and boring.............	$13.00
12″ Lathe Blocks to swing 15″ for turning and boring.............	15.00
13″ Lathe Blocks to swing 18″ for turning and boring.............	15.00
15″ Lathe Blocks to swing 20″ for turning and boring.............	17.00
16″ Lathe Blocks to swing 22″ for turning and boring.............	20.00
18″ Lathe Blocks to swing 24″ for turning and boring.............	25.00

HOW TO TEMPER A LATHE TOOL

A forged lathe tool is made of crucible or tool steel. In tempering a lathe tool care should be taken so as to heat it slowly and evenly for a distance of about 2 inches from the cutting point. When the heated part becomes a cherry red, immerse the tool in cold water about 1½ inch from the cutting point. (See drawing.) The heat is now driven into the shank of the tool, so as to assist in drawing the temper. After the point has become cool, remove the tool from the water and polish the hardened surface with a piece of emery cloth, then wipe this surface with an oily rag. The heat in the shank of the tool will now drive the temper toward the cutting edge. When this edge becomes a brown straw color immerse the entire tool in cold water.

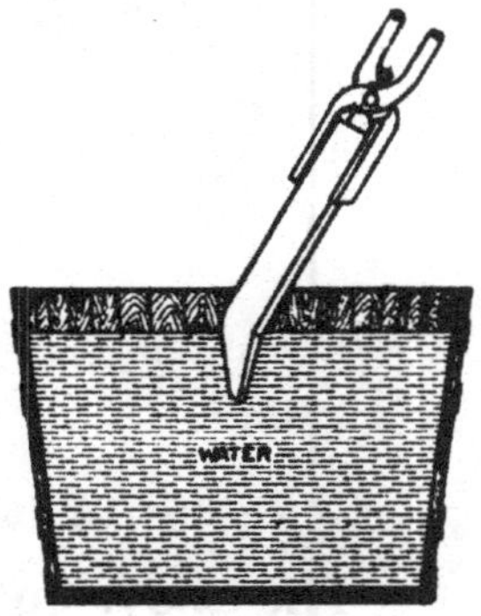

HOW TO ANNEAL A PIECE OF TOOL STEEL

Heat the steel slowly and evenly to a dark red. Then place in box of lime or ashes, cover completely and let remain over night. In the morning the piece will be annealed ready for machining.

To water-anneal a piece of tool steel, heat slowly and evenly until a dark red. Then hold in the tongs in a shaded corner until all color has left. Place a small pine stick against the steel. When the steel is cool enough so that it will not smoke the pine stick, then immerse the steel quickly into cold water and it is ready for machining.

CASE HARDENING

To case harden a piece of machinery steel, for example, a ball race: Heat the ball race in an even fire until the piece becomes a dark cherry red, take it out and with a small piece of cyanide of potash in one hand touch the heated part where you wish it case hardened. The cyanide will dissolve slowly and be absorbed by the ball race. After the surface to be hardened has received a thorough coat of cyanide, place the ball race back in the fire and heat slowly for about one minute so that it will thoroughly absorb the cyanide, remove the piece from the fire and plunge into cold water.

USING A REAMER IN THE LATHE

Fig. 71 shows a practical method of using a reamer in the lathe so that it will not draw or bite. Attach a dog to the reamer, place the end of the reamer on the tail center and with a tool in the tool post, feed the compound rest up so that the dog will be supported by the top of the rest. The reamer is held back on the center by the tool in the tool post, which presses against the dog. Start the lathe and feed the reamer into the work by the aid of the tail stock wheel. The reamer will now feed evenly, and drive the carriage before it, and the weight of the carriage will prevent the reamer from gripping.

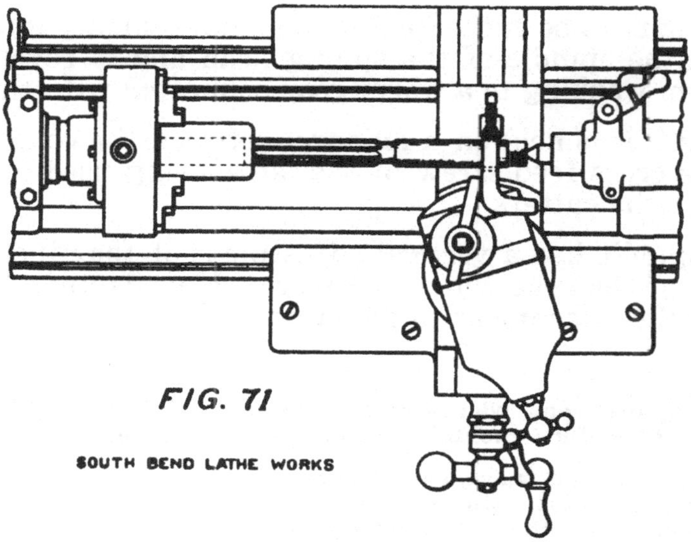

FIG. 71

SOUTH BEND LATHE WORKS

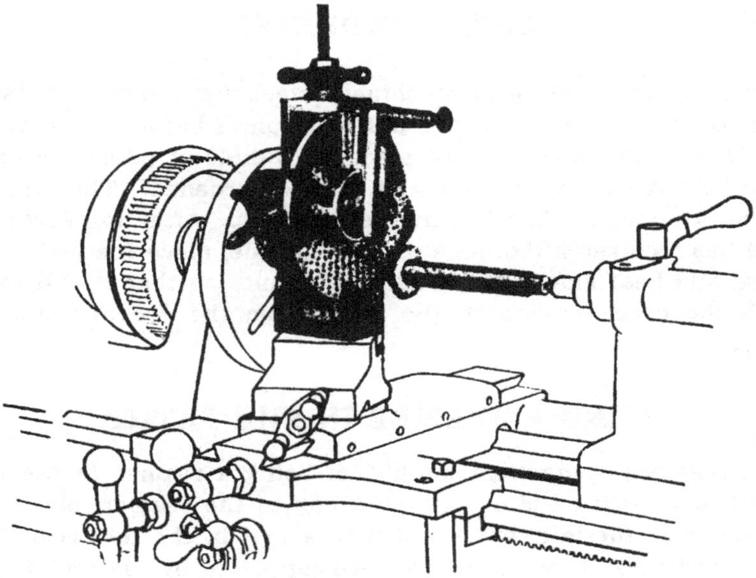

No. 1 GEAR CUTTING ATTACHMENT FOR SOUTH BEND LATHES

No. 1 Gear Cutting Attachment is intended for cutting small light gears on the lathe.

The blank to be cut is held on a mandrel fitting into a taper socket in the spindle of the sliding head and is revolved by a steel worm working in a worm gear connected with the dials.

The cutter is held on a mandrel between the centers of the lathe, the cross feed screw of the lathe being used to feed the work over the cutters.

The spindle has a vertical adjustment of four inches. The diameter of the gear which can be cut depends upon the application of the attachment to the lathe.

The beginner who wishes to become a first class machinist should observe the three fundamental rules of the shop, as follows:

1. **Take your measurements accurately.**
2. **Keep your tools sharp.**
3. **Keep your lathe well oiled and clean.**

INFORMATION ON GEARS

Diameter, when applied to gears, is always understood to mean the pitch diameter.

Diametral Pitch is the number of teeth to each inch of the pitch diameter.

Example: If a gear has 40 teeth and the pitch diameter is 4 inches, there are 10 teeth to each inch of the pitch diameter and the diametral pitch is 10, or in other words, the gear is 10 diametral pitch.

Number of Teeth required, pitch diameter and diametral pitch given. Multiply the pitch diameter by the diametral pitch.

Example: If the diameter of the pitch circle is 10 inches and the diametral pitch is 4, multiply 10 by 4 and the product, 40, will be the number of teeth in the gear.

Number of Teeth required, outside diameter and diametral pitch given. Multiply the outside diameter by the diametral pitch and subtract 2.

Example: If the whole diameter is 10½ and the diametral pitch is 4, multiply 10½ by 4 and the product, 42, less 2, or 40, is the number of teeth.

Pitch Diameter required, number of teeth and diametral pitch given. Divide the number of teeth by the diametral pitch.

Example: If the number of teeth is 40 and the diametral pitch is 4, divide 40 by 4 and the quotient, 10, is the pitch diameter.

Outside Diameter or size of gear blank required, number of teeth and diametral pitch given. Add 2 to the number of teeth and divide by the diametral pitch.

Example: If the number of teeth is 40 and the diametral pitch is 4, add 2 to the 40, making 42, and divide by 4; the quotient, 10½, is the whole diameter of gear or blank.

Distance Between Centers of two gears required. Add the number of teeth together and divide one-half the sum by the diametral pitch.

Example: If the two gears have 50 and 30 teeth, respectively, and are 5 pitch, add 50 and 30, making 80, divide by 2, and then divide the quotient, 40, by the diametral pitch, 5, and the result, 8 inches, is the center distance.

THE CUTTING SPEED FOR DIFFERENT METALS

The following cutting speed is recommended where high speed cutting tools are used:

Cast Iron	55 feet per minute
Machine Steel	35 " " "
Wrought Iron	40 " " "
Tool Steel, Annealed	25 " " "
Brass	100 " " "
Bronze	70 " " "
Grey or Red Fiber	50 " " "

RULE FOR GEARING UP ENGINE LATHES FOR SCREW CUTTING

Multiply both the number of threads per inch on the lead screw and the number of threads per inch on the bolt to be cut by any number that will give for a product gears that you have with the lathe.

Example: To cut eleven threads per inch; lead screw on lathe, 8 threads per inch; multiply by any number, say 4:

4 times 8 are 32.
4 times 11 are 44.

If the 32 and 44 gears are not in the lathe equipment try multiplying by 5:

5 times 8 are 40.
5 times 11 are 55.

If the thread to be cut is finer than the thread in the lead screw the smaller gear goes on the spindle, while the larger gear goes on the lead screw.

Always measure the thread when you have taken the first chip to be sure you have made no mistake.

For finding the pitch of screw, see page 29.

Hinge Guard Partly Open

Hinge Guard Closed

GEAR GUARDS FOR SOUTH BEND LATHE

The illustrations above show the general appearance of South Bend lathe with gear guards attached. Fixed guards cover the back gears, and a hinged guard covers all end gears. This guard is adjustable, and may be swung open when necessary. Gear guards are made of cast iron.

HOW TO ANNEAL BRASS OR COPPER

In working brass and copper, it will become hard, and if hammered to any great extent will split. To prevent cracking or splitting, the piece must be heated to a dull red heat and plunged into cold water; this will soften it so it can be worked easily. Be careful not to heat brass too hot, or it will fall to pieces. The piece must be annealed frequently during the process of hammering.

HOW TO BRAZE

Clean the article thoroughly, and better to polish with sand paper. Fasten the parts to be brazed firmly together, so they will not part when heated in the fire. Place over a slow fire of charcoal or well coked coal. Place on the parts to be brazed a small quantity of pulverized borax; as soon as this is done boiling and has flowed to all parts, then put on the spelter; when the spelter melts it will generally run in globules or shot. Jar the piece by gently striking with a small piece of wire; this will cause the spelter to flow to all parts.

Fig. 301

Semi-Machined Chuck Plate

Fig. 300

Rear View of Lathe Chuck

FITTING CHUCKS TO THE LATHE

Fig. 301 show a Cast Iron Semi-Machined Chuck Plate. It is called "semi-machined" because it has been bored, faced, and threaded to fit the spindle nose of the lathe it is intended for.

To fit a lathe chuck to the spindle, select a semi-machined chuck plate with the flange large enough so that it may be turned to fit the recess in the chuck it is intended for. Screw this chuck plate on the spindle nose so that the hub of the plate fits up against the shoulder of spindle. Machine the chuck plate all over and turn the diameter of the flange to fit the recess of chuck, shown in Fig. 300. The fit should be accurate, — not tight nor loose, but just right. Remove the chuck plate from the spindle, drill the necessary holes in plate, 1-16″ larger than screw, attach it to the chuck with proper screws. Chuck is fitted ready for use.

When ordering a lathe and you wish a chuck included, always specify that the chuck be fitted to the lathe complete with chuck plate before the lathe leaves factory. The lathe manufacturer has suitable equipment for machining and fitting lathe chucks to lathes of their own make, charging approximately the actual cost of the labor and material.

SIZE OF LATHE CHUCKS FOR A LATHE

We show herewith a list giving the size of lathe chuck most suitable for the various size lathes. Of course it may be necessary to depart from this list for special work but for general all-around work in the machine shop these sizes will be found practical.

11-inch Lathe, size of chuck... 4″ to 6″ inclusive
12-inch Lathe, size of chuck... 4″ to 8″ inclusive
13-inch Lathe, size of chuck... 5″ to 9″ inclusive
15-inch Lathe, size of chuck... 6″ to 10″ inclusive
16-inch Lathe, size of chuck... 6″ to 12″ inclusive
18-inch Lathe, size of chuck... 8″ to 14″ inclusive

METRIC THREADS ON AN ENGLISH LEAD SCREW

To cut Metric Threads on a South Bend Lathe using a regular standard English lead screw, use the Compound Idler or connecting gears 50 and 127, the No. 127 Gear to mesh with spindle stud. Use an idler to connect the 50-tooth gear with Gear on Lead Screw.

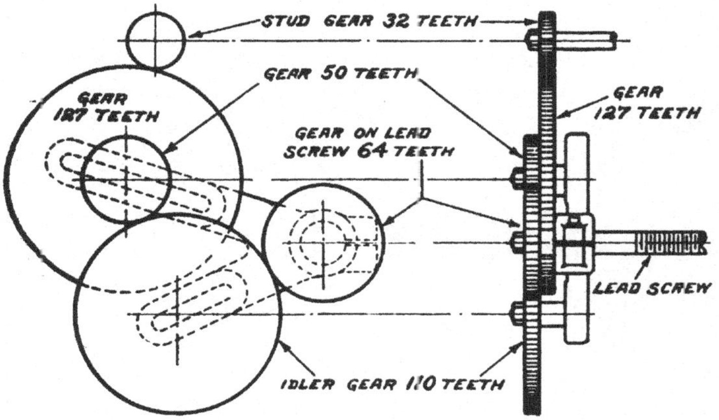

Arrangement of gearing to cut 16 thread per centimeter on a No. 34 South Bend lathe.

When Metric Threads are to be cut as illustrated the regular Index Chart of lathe may be used in selecting gears for the different pitches. Read the chart as so many threads per centimeter, instead of so many threads per inch.

South Bend lathes may be equipped with a Metric Lead Screw, instead of an English Lead Screw, if desired.

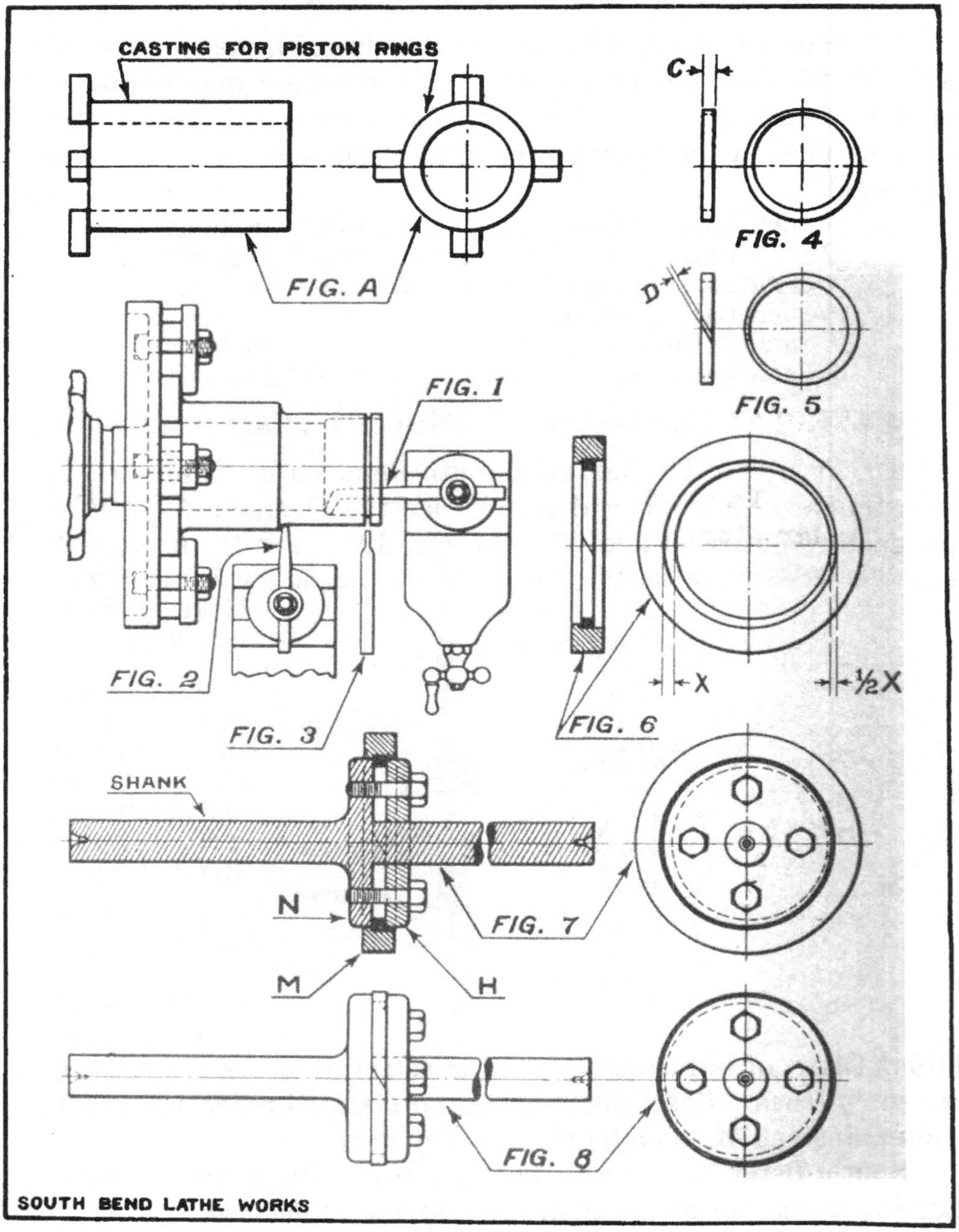

Operations on a Piston Ring

MAKING A PISTON RING

Fig. A—Shows cast iron drum for piston rings. Most foundries carry these drums in stock.

Fig. 1—Shows first operation; casting is clamped to face plate of lathe and bored the required diameter.

Fig. 2—Shows turning operation; after boring, set casting off center so that the thickness of wall on a line running directly through the center diameter, will be twice as thick on the one side as on the other. (See Fig. 6.)

Fig. 3—Shows cutting-off operation.

Fig. 4—Shows a ring, after it has been bored, turned and cut off.

Fig. 5—Shows ring sawed on a slant at the thinnest side. The slant joint must fit tight when fitted into ring gauge.

Fig. 6—Shows piston ring fitted to ring gauge. The diameter of the hole in this gauge is about 1/32nd larger than the cylinder bore, so when ring gauge and ring are fastened on arbor, the ring gauge may be removed, and there will be extra stock to be machined on outside diameter of the piston ring, to conform to the bore of cylinder.

Piston ring when finished should be two-thousandths less in diameter than the bore of cylinder.

Fig. 7—Shows arbor for turning outside diameter of piston ring. The piston ring is fitted to ring gauge, (Fig. 6), so that the slot "D" (Fig. 5) is made a tight joint, then put ring gauge and ring on arbor; slip the ring gauge on the arbor at "N," this makes the outside of the piston ring concentric with the arbor.

Fasten flange "H" with the four screws; remove ring gauge in the direction of shank of arbor, and piston ring is ready to be finished on outside diameter.

Fig. 8—Shows piston ring on arbor finished. The thickness of the piston ring at "X" (Fig. 6) should be equal to the depth of the slot in piston. The width of the piston ring at "C" (Fig. 4) should be a perfect sliding fit in this slot.

The diameter of the piston ring hole before it is sawed has to be found by experiment, according to the size of the piston the ring is to be fitted on. If, however, you are replacing a piston ring on a piston, measure the inside of hole in old piston ring, as this will give you a definite size.

Regular Equipment, as Illustrated Under Lathe, is Included in the Price

No. 34—13-INCH SWING SOUTH BEND SCREW LATHE

Fitted with Automatic Longitudinal Feed and Power Cross Feed

The No. 34 Lathe illustrated above is surpassed by none for general all-around work in the machine shop, manual training school and general repair shop. It has a great many practical features that makes it valuable in manufacturing and which enables it to take care of the various jobs that come up.

Lathe and regular equipment:

1—No. 34 South Bend Lathe, 13″ swing, 5-foot bed, complete with Improved Reverse on head, graduated compound rest, automatic longitudinal feed, automatic cross feed, large and small face plates, two steel centers, center rest, change gears and adjustable stop for screw cutting, necessary wrenches and double friction countershaft.

Price F. O. B. South Bend.........................**$165.00**

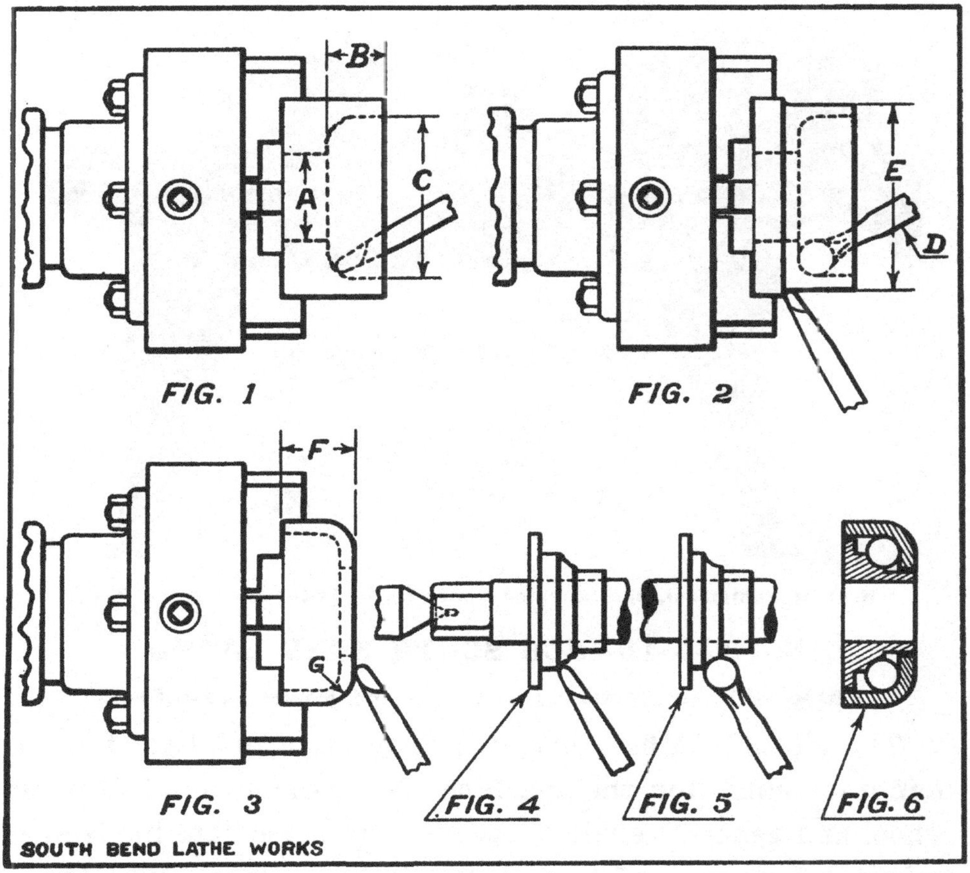

MAKING A BALL RACE AND CONE

Fig. 1—Shows a piece of steel held in the chuck, being bored and machined for making the cone.

Fig. 2—Shows a second operation, turning for the Ball Race and also turning the outside of the cone.

Fig. 3—Shows the last operation. The cone is being faced on the end and the corners rounded. All that is now necessary is to case harden it.

Fig. 4—Shows a Ball Race being machined on centers on an arbor. The Ball Race is first chucked and bored similar to the first operation on the cone.

Fig. 5—Shows the Ball Race on an arbor between centers. The Ball Race proper being machined by a form tool.

Fig. 6—Shows the Ball Race and cone assembled complete.

The material that a Ball Race and Cone are made of is Machinery Steel. Crucible Steel may be used, but Machinery Steel, case hardened, will give excellent results. For case hardening see page 49.

Regular Equipment, as Illustrated Under Lathe, is Included in the Price

No. 37—15-INCH SOUTH BEND LATHE

Fitted with Automatic Longitudinal and Power Cross Feeds

The No. 37 Lathe illustrated above is a practical tool for manufacturing, for the machine shop, the manual training school and general repair shop. It is intermediate in size and weight between the 13″ and 16″ lathes. This 15″ lathe is capable of taking care of a great variety of work.

Lathe and regular equipment:

1—No. 37 South Bend lathe, 15″ swing, 6-foot bed, complete with Improved Reverse on head, graduated compound rest, automatic longitudinal feed, automatic cross feed, two steel centers, center rest, follower rest, change gears and adjustable stop for screw cutting, necessary wrenches and double friction countershaft.

Price F. O. B. South Bend..........................$215.00

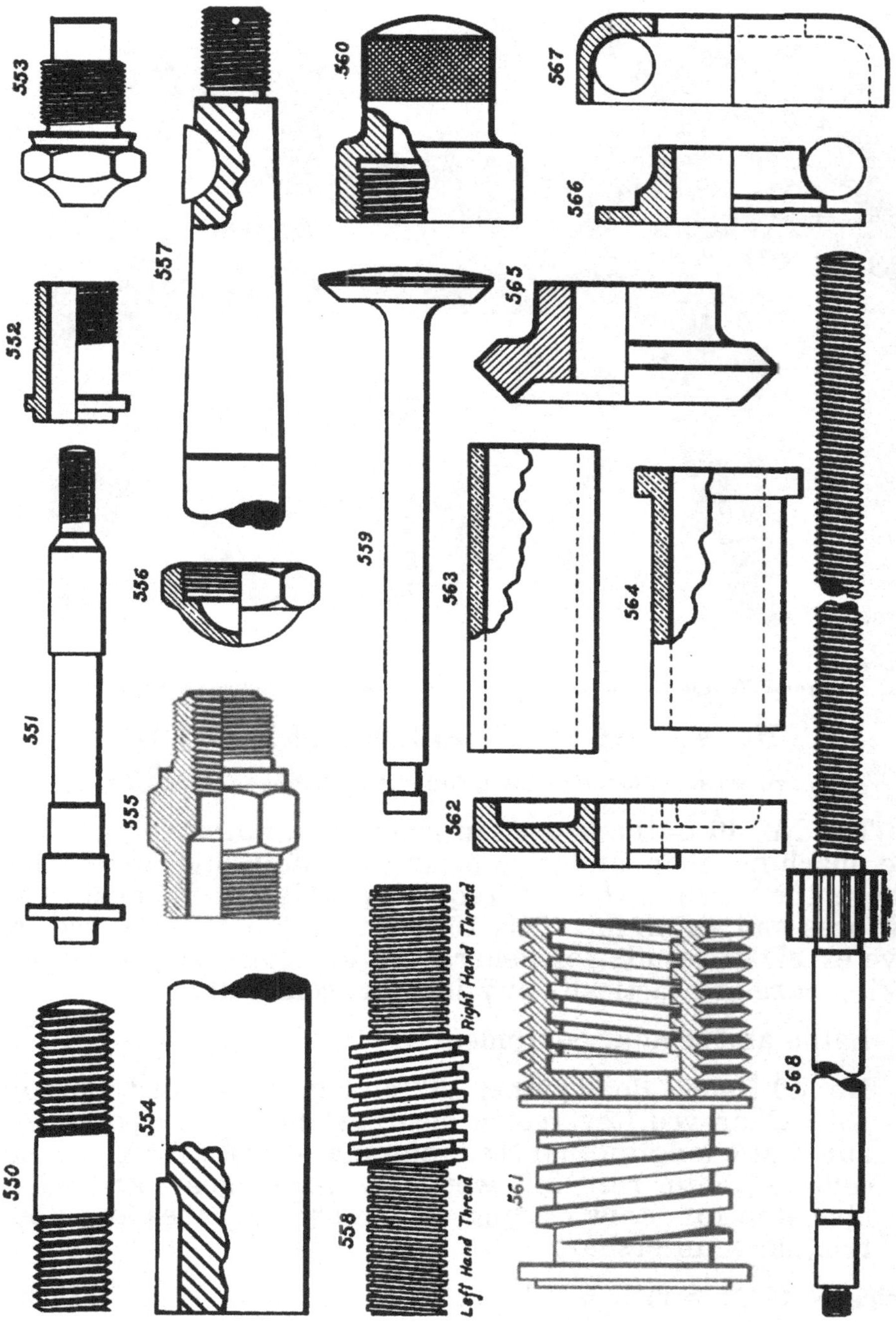

The Cuts Show Only a Few Examples of Lathe Work Produced on a South Bend Lathe

Regular Equipment, as Illustrated Under Lathe, is Included in the Price

No. 40—16-INCH SOUTH BEND LATHE

Fitted with Automatic Longitudinal and Power Cross Feeds

The No. 40 Lathe is used in general manufacturing, also in the machine shop and the practical all-around repair shop where the work is heavy, requiring a lathe of considerable stiffness and strength. This 16″ lathe in 8 or 10-foot bed, will give excellent results for general work because it is capable of taking care of practically any job that comes up.

Lathe and regular equipment:

1—No. 40 South Bend lathe, 16″ swing, 8-foot bed, complete with Improved Reverse on head, graduated compound rest, automatic longitudinal feed, automatic cross feed, two steel centers, center rest, follower rest, change gears and adjustable stop for screw cutting, necessary wrenches and double friction countershaft.

Price F. O. B. South Bend........................$282.00

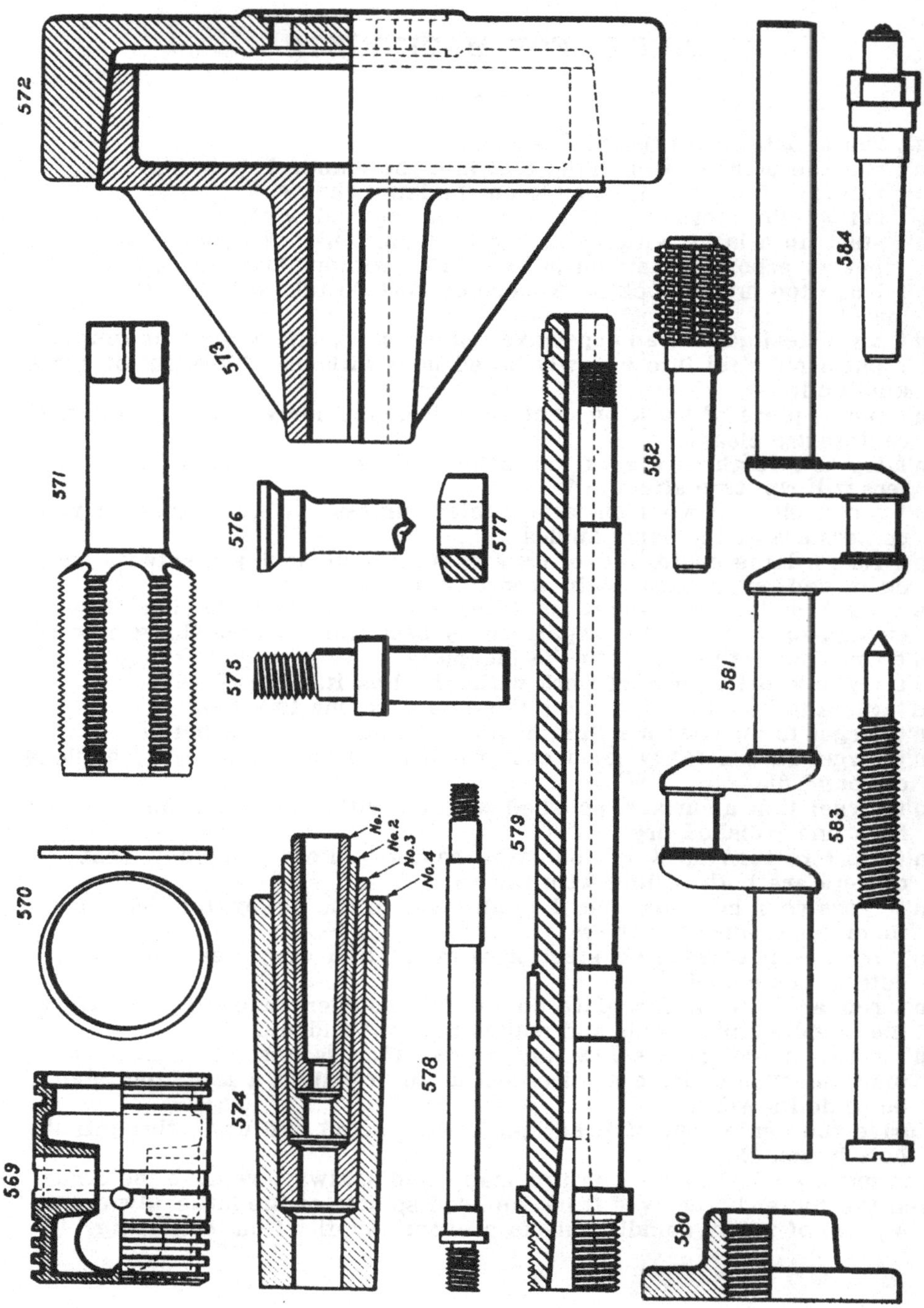

The Cuts Show Only a Few of the Many Pieces That Can be Produced on a South Bend Lathe

DON'TS FOR MACHINISTS

From "Machinery"

Don't run a Lathe with the belt too loose.
Don't run the point of your lathe tool into the mandrel.
Don't rap the chips out of your file on the lathe shears.
Don't set a lathe tool below the center for external work.
Don't start up a lathe without seeing that the tailstock spindle is locked.
Don't put an arbor or shaft on lathe centers without lubricant on them.
Don't leave too much stock on a piece of work to take off with the finishing cut.
Don't try a steel gage or an expensive caliper on a shaft while it is running.
Don't put a mandrel into a newly bored hole without a lubricant of some kind on it.
Don't put a piece of work on centers unless you know that the internal centers are clean.
Don't try to straighten a shaft on lathe centers, and expect that the centers will run true afterwards.
Don't put a piece of work on lathe centers unless you know that all your centers are at the same angles.
Don't take a lathe center out of its socket without having a witness mark on it, and put it back again according to the mark.
Don't start polishing a shaft on lathe centers without having it loose enough to allow for the expansion by heat from the polishing process.
Don't run your lathe tool into the faceplate.
Don't try to knurl a piece of work without oiling it.
Don't run a lathe an instant after the center begins to squeal.
Don't forget to oil your machine every morning; it works better.
Don't forget that a fairly good center-punch may be made from a piece of round file.
Don't forget that a surface, polished with oil will keep clean much longer than one polished dry.
Don't start to turn up a job on lathe centers unless you know that the centers are both in line with the ways.
Don't cross your belt laces on the side next to the pulley, for that makes them cut themselves in two.
Don't try to cut threads on steel or wrought iron dry; use lard oil or a cutting compound.
Don't run a chuck or faceplate up to the shoulder suddenly; it strains the spindle and threads and makes removal difficult.
Don't screw a tool post screw any tighter than is absolutely necessary; many mechanics have a false idea as to how tight a lathe tool should be to do its work.
To drive the center out of head spindle use a rod and drive through the hole in spindle.
When putting a lathe chuck on the head spindle, always remove the center.
When the center is removed from the head spindle of the lathe, always put a piece of rag in spindle hole to prevent any dirt from collecting.

Made in United States
North Haven, CT
10 November 2026

22106116R00043